Data Intensive Industrial Asset Management

Farhad Balali • Jessie Nouri • Adel Nasiri
Tian Zhao

Data Intensive Industrial Asset Management

IoT-based Algorithms and Implementation

 Springer

Farhad Balali
University of Wisconsin–Milwaukee
Milwaukee, WI, USA

Jessie Nouri
University of Wisconsin–Milwaukee
Milwaukee, WI, USA

Adel Nasiri
University of Wisconsin–Milwaukee
Milwaukee, WI, USA

Tian Zhao
University of Wisconsin–Milwaukee
Milwaukee, WI, USA

ISBN 978-3-030-35929-4   ISBN 978-3-030-35930-0   (eBook)
https://doi.org/10.1007/978-3-030-35930-0

This Springer imprint is published by the registered company Springer Nature Switzerland AG
The registered company address is: Gewerbestrasse 11, 6330 Cham, Switzerland

To Sarah, the love of my life.

—Adel

To my beloved parents and sister for their endless support, encouragement and sacrifices.

—Farhad

To my parents, for encouraging me to be brave and going after what I believe in, and to my sisters, for giving me endless love in taking my journey.

—Jessie

Preface

The concept of digital transformation is penetrating all aspects of our life. This transformation has been fueled by fast technological growth in computational and communication components and systems. A major evolving impact has been on all sectors of industrial systems in which traditional problems are being formulated to be solved with digital technologies. There are numerous examples on early adopters who have embraced digital technologies and have become successful in their business. On the contrary, in many instances, companies that fell behind the digitalization trend either disappeared or lost significant market share to competitors. The concept of the Internet of Things (IoT) can be considered a subset of digital transformation in which every device is connected to the Internet and can provide data. The devices include a range of components from a wall thermostat to a vending machine and to large industrial machines and facilities. By some estimates, 26 billion devices will be connected to the Internet by 2020. This is a major shift in way of life, information gathering and sharing, decision-making, and planning. This revolution is propelled by several technological advancements in recent years. As a starter, embedded systems, which connect hardware and software, have experienced a major leap in increased local processing power, low overhead Windows and Linux operating systems that can run on a chip, and also on development and deployment of low overhead data protocols. Examples of these protocols that can be run on low-cost embedded boards include Message Queuing Telemetry Transport (MQTT) and Advanced Message Queuing Protocol (AMQP). Another major development in recent years is the development of faster and lower-power wireless communications. 1G enabled voice transmission, 2G enabled texting, 3G enabled video transfer, and 4G enabled video streaming. 5G promises to revolutionize data transmission by providing higher speed and bandwidth to connect every device to the Internet. The last component is the significant advancement in development and deployment of cloud computing. Data from any system, machine, plant, or factory can be sent to data network and cloud gateway to use the enormous computation power and cloud functions.

The IoT has found many applications in industry, healthcare, smart cities, farming, telecommunications, and water and energy systems. Broader benefits of the IoT include reduced need for human efforts, improved productivity and reliability, lowered capital cost, better customer experience, and real-time marketing. According to some estimates, IoT expenditure will surpass $1 trillion by 2022 with a large year over year growth. Industrial IoT (IIoT) is the application of IoT in industrial processes and is projected to absorb a significant portion of the IoT investment and development. Manufacturing, industrial asset management, and connected products are considered as major elements of the IIoT.

Data from every device and every component of a machine within a factory or a manufacturing plant can be collected through a data network and transferred to cloud platform for analytics and cloud functions. In the process and at every stage of sensor, component, machine, plant line, and factory edge, raw data is reduced and is converted to information. In the cloud stage, various cloud functions such as streaming analytics, data storage, data analytics, artificial intelligence and machine learning, and visualization can be applied. In addition, the data layer in cloud can be connected to manufacturing execution system (MES) and enterprise resource planning (ERP) to create a connected system from devices and machines to data layer and business later. This connected system offers greater productivity, efficiency, and safety.

The same concept can be applied for industrial assets to enable data-driven industrial asset management. For this approach, four levels of maturity can be defined. In the first level, there is no data connection, and only regular maintenance is performed according to manufacturer's recommendation. In the second level, sensors are connected to assets, and reactive support, troubleshooting, and maintaining are conducted. At the third level, data are collected from sensors and analyzed to provide various benefits including data contextualization, proactive maintenance, smarter decision-making, decreased cost, and increased productivity. At the last and ultimate level, data is centralized, and artificial intelligence is applied to integrate supply chain, perform prescriptive maintenance, reduce risk, change behavior, and transform the business. This book has a major focus on data-driven industrial asset management and is intended to serve as a guideline for researchers, students, instructors, and the industry. It specifies the challenges raised by IoT devices and describes the challenges of parameter selection, statistical data analysis, predictive algorithms, big data storage and selection, data pattern recognition, machine learning techniques, asset failure distribution estimation, reliability and availability enhancement, condition-based maintenance policy, failure detection, data-driven optimization algorithm, and multi-objective optimization approach.

Milwaukee, WI, USA   Farhad Balali
Milwaukee, WI, USA   Jessie Nouri
Milwaukee, WI, USA   Adel Nasiri
Milwaukee, WI, USA   Tian Zhao

Contents

List of Figures

List of Tables

Author Biography

Farhad Balali obtained his Ph.D. from the University of Wisconsin-Milwaukee in December 2019. He is also a Senior Control Algorithm Engineer in Advanced Development and Application Department at the Johnson Controls Co. Farhad was born in Tehran, Iran, and studied Industrial Engineering at K.N. Toosi University of Technology. He obtained his Master's degree in Industrial Engineering from the University of Wisconsin-Milwaukee in December 2015. He started working in the Center for Sustainable Electrical Energy Systems during his Master's program since 2014. His Master's thesis titled "An Economical Model Development for a Hybrid System of Grid Connected Solar PV and Electrical Storage System." His current research interests are predictive models, asset management, degradation models, reliability assessment, data-driven algorithms, machine learning (ML), artificial intelligence (AI), Internet of Things (IoT), and statistical data analysis. His Ph.D. dissertation is titled "A Data-Driven Predictive Model of Asset Management for Distributed Electrical Systems." Farhad is a member of the Editorial Board for *Internet of Things and Cloud Computing* and *Progress in Energy & Fuels* journals. During his graduate program, he published six journal papers, three conference papers, and one book chapter. Furthermore, he served as a reviewer for several prestigious journals such as *Renewable Energy*, *Energy*, *International Journal of Energy Research*, *Clean Technologies and Environmental Policy*, and *Journal of Industrial Engineering International* and for IEEE Energy Conversion Congress and Exposition (ECCE) and International Conference on Energy Engineering and Environmental Protection.

Chapter 1
Internet of Things (IoT): Principles and Framework

1.1 Internet of Things (IoT)

"Internet of Things" (IoT) refers to a connected network of the smart devices, which are able to communicate through a wired or wireless communication network. IoT is also called the Internet of Everything (IoE) that represents new concepts including smart interactions between machines and devices [1]. As Bauer et al. (2014) stated in [2], the "Industrial Internet of Things" (IIoT), which refers to a subgroup of comprehensive IoT, is defined as the "real-time capable, intelligent, horizontal, and vertical connection of people, machines, objects, and Information and Communications Technologies (ICT) systems to dynamically manage complex systems." "Seamless integration of physical objects such as sensors or home appliances (i.e., things) and services" through a communication network has been presented by De Leusse et al. (2009) [3]. In Europe, the term "Industry 4.0" is mostly used in reference to IIoT as proposed in 2011, for large-scale production projects [4].

A smart connected IoT system is a highly automated network including the components connected to the cyber physical system and gateways using the latest communication technologies with the help of the Internet. What brings the benefit to an IoT network is the connection between many smart devices and units, which are able to communicate seamlessly and efficiently to measure, collect, and analyze required data. The data is then used as a raw input to feed the developed algorithms to make the system as smart as possible and to be able to take the necessary actions with the minimum amount of human intervention [5].

The entire value of the IoT highly depends on the interaction between the units. Nowadays, many of the objects surrounding us are parts of an IoT paradigm in one form or another. A considerable portion of the world's population is connected to each other through the Internet. In a smart connected IoT system, this connection applies to the device-to-device communication with a minimum amount of human intervention. The Internet is playing a vital role in a smart connected system, and without the Internet, the interconnection between the smart units of the network is

© Springer Nature Switzerland AG 2020
F. Balali et al., *Data Intensive Industrial Asset Management*,
https://doi.org/10.1007/978-3-030-35930-0_1

Fig. 1.1 Main elements of a
smart connected system

not possible. Existing Internet standards and protocols make the system able to work with various types of hardware and software to collect information, perform data analytics, and take action at the right time and right place. However, fully integrated future of the Internet includes various communication networks such as Bluetooth, RFID, Wi-Fi, telephonic data services using the sensor and actuator nodes, and gateways [6].

Figure 1.1 represents the main elements of a smart connected system. The arrows are the key benefit of a smart connected system, which is referred to as the "network" element. The network is the interconnection between the physical and cyber systems. "Machine and process data" includes any physical devices installed in the system, which has the responsibility of collecting data and communicating with other devices as well as with the control center. "Control center" is in charge of regulating these interconnections between the different cyber and physical points of the network to make sure all the sectors are working under their standard and normal conditions. Standardized data will be processed using "data analytics" techniques, and useful information will be extracted. "Artificial intelligence (AI)" principles use the output of the "data analytics" to train the "control center." AI can make the control center smart enough to work as a strong human brain and make best decisions at the right time with minimal human interventions. The "business enterprise" is in charge of solving large-scale problems using the AI output. And last but not least, the knowledge extracted from all previous actions will be stored in the "cloud" for further utilization.

To summarize, IoT harvests the collected data from the smart devices and performs analytics to evaluate the performance of the units, assets, and services.

The outcomes of the analyses can be identifying any abnormality or pattern and monitoring of the system status as close to real time as possible [7]. Dataflow works as the blood of the network to keep it alive and healthy. Data analytics can be considered as a blood test, or any related test for health monitoring, to detect malfunctions in any part of a smart connected system. Business environments interconnect the virtual and the real worlds, relating companies and individuals in different roles, e.g., module providers, machine-to-machine (M2M) service providers, network operators, and users, that interact and share connected hardware, software, and platforms with one another [8–10].

Smart connected systems exponentially expand the opportunities and improve the control measures such as availability, reliability, efficiency, utilization, capabilities, functionality, maintainability, and value chain of the products and services within the whole supply chain of the firms. For instance, IoT implementation is improving the "reliability" through the advanced monitoring techniques. As an example, the traditional asset management techniques use only maintenance policies prescribed by the manufacturer. These types of approaches are not optimal as they do not take the current working condition and history of the asset and its utilization into account. At the same time, modern asset management is based on the real-time condition of the assets to minimize failures and unexpected shutdowns. Consequently, the latter approach reaches a higher level of asset reliability obtained through application of IoT and smart devices [11, 12]. Therefore, IoT implementation can help businesses to optimize their control measures.

To that end, during the last couple of decades, IoT principles have been successfully applied to different sectors such as healthcare, transportation, manufacturing, agriculture, energy, etc. Indeed, "things" can apply to many different technologies in various sectors, while the core of the connected system is still the smart communications between the units. IoT abilities can be added to the commonly used objects to make them smart and ready to interact with other devices [13]. IoT can enhance the interactions between the units as well as between human and devices. For instance, the relationship between the seller and customer can be reshaped using digitized communication services [14].

Companies invest in the IoT area for different purposes such as fully automating the system, minimizing cost, increasing safety for the personnel, enhancing supply efficiency, reducing failures, increasing customer satisfaction, and optimizing guarantee policies [15]. Since the competition is ongoing among companies to become leaders in the market in their sector and IoT is a huge competitive advantage, there are a lot of companies which are trying to adopt the IoT principles to achieve benefits and to increase their revenue through services. Although the concept can offer many benefits, companies are mostly uncertain on how return on investment can be formulated for the high cost of implementation. It is crucial for companies to carefully review the outcomes, benefits, and challenges before implementing an IoT platform [16]. IoT has the potential to become a powerful competitive advantage for many types of businesses, which can carefully identify the outcomes and challenges [17].

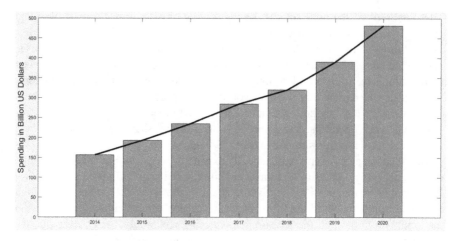

Fig. 1.2 Projected IoT spending worldwide [19]

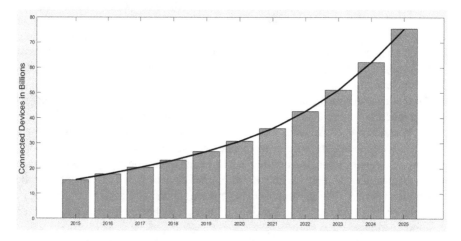

Fig. 1.3 IoT connected devices installed base worldwide from 2015 to 2025 (in billions) [20]

Smart devices may significantly affect the infrastructure of the businesses and force them to redesign, rethink, and retool their properties. All these changes completely reform the nature of some industries and also change the meaning of the competition [18]. Competition is not only limited to the products or services provided by industries. For instance, in the process of shopping for an item, the origin and types of the raw materials, how and where it is manufactured, current functionality and performance, future services, and recyclability might change a customer's final decision. Among the similar competitor products, the customer might pick the one which is the best in the overall value chain.

Implementing IoT infrastructure comes with challenges. Figures 1.2 and 1.3, respectively, are representing the data provided by "Statista" [19, 20] on both the spending worldwide budget (in billion US dollars) on IoT projects and the IoT connected devices forecasted to be installed worldwide from 2015 to 2025. An increasing trend in both graphs proves the wide acceptance popularity of the IoT applications in public and private entities [21, 22] . However, more connected smart devices bring more complexity to the system. As the number of IoT smart devices increases, the size and dimensions of the collected data grow exponentially. Therefore, one of the main upcoming challenges would be handling the enormous size of the data and extracting the most beneficial information from the collected data points as fast as possible. Sometimes, more data points only add more complexity to the system, while the same benefit could be achieved using smaller portions of the data.

At the same time, coupling smart IoT devices increases the overhead cost. The associated investment cost of the infrastructure is considerable compared with the cost of deploying simple smart devices such as sensors. Data collection, correction, storage, and analysis are all among the cost-intensive tasks of building a network. The whole data process should have economic justifications to make a project more feasible.

The wide applications of the IoT significantly affect the whole supply and value chains of the firms and business as well as the routine life of the individuals. Some believe that the IoT "changes everything" surrounding us. The simplest example of this can be attributed to using smartphones. A lot of people are using smartphones and indeed are close to a very smart unit. For instance, based on one's activities, the smartphone is able to predict driving routes for different times of the day. Usually, it can be between one's home and work place with possible stops that are made frequently. A smartphone is able to warn the user of any abnormal traffic or accident, special events such as carnivals, special games, concerts, etc. Another everyday example of smart devices can be smartwatches. A smartwatch can be programmed to predict any unhealthy condition of body and call relatives or emergency services. As it is evident from these two examples, smart devices are trying to automate everything, enhance the quality of life, and ultimately change the way of life.

There are also numerous industrial applications for IoT. Industrial IoT (IIoT) is the application of IoT in industrial processes. Some instances of IoT applications in a smart city are as depicted in Fig. 1.4. An example of the IoT applications in transportation is hyper-local decisions on arranging road crews in severe winter weathers using IoT platforms to save lives as well as to minimize the maintenance cost. Another example can be the surveillance services offered to air traffic control providers around the world by advanced smart trackers for the airplanes. IoT has also had a large impact on the healthcare sector. Tracking patients, staff, and inventory and monitoring patient's conditions online and pills containing microscopic sensors are some examples of how IoT applications transformed healthcare services.

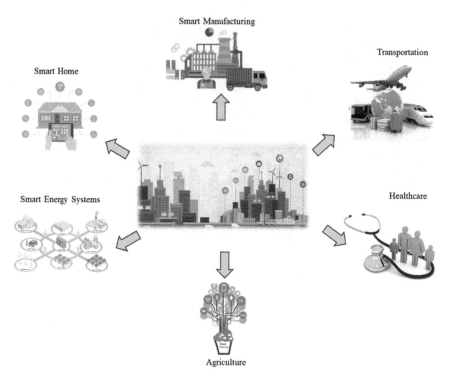

Fig. 1.4 Major applications of the Internet of Things (IoT) in a smart city

1.2   Smart Systems vs. Smart Connected Systems

During the 1960s and 1970s, the potential roles of the "information technology" (IT) significantly reshaped the businesses and added more value concepts to the competitive landscape. "Automated individual decision-making" and "computer-aided design" (CAD) are among the first examples, which received much attention during those periods [24] and formed a foundation for creating smart systems. Smart systems significantly increase the efficiency and productivity of the processes due to the available data allowing decision-makers to gain information and make decisions accordingly.

The connection capability between smart systems was added during the 1980s and 1990s with the emergence of the Internet. It was proven to businesses that the benefits of using Internet and data networks far overweigh the cost of implementation and usage. Smart devices were integrated into traditional systems to monitor and control the process; however, some systems were very outdated and did not have the capability to turn into a smart connected system.

The focus of this section is mostly on the differences between smart systems and smart connected systems. Figure 1.5 is a graphical representation of this difference. It should be noted that using smart devices is not a new phenomenon although it has

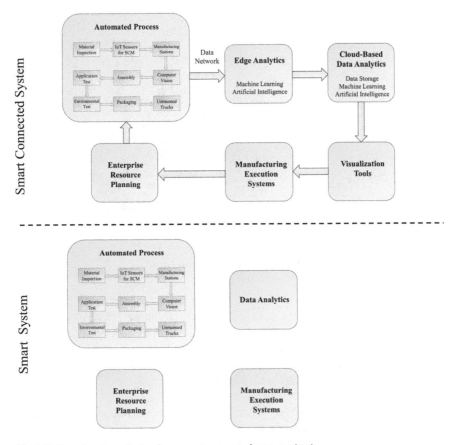

Fig. 1.5 Smart systems (bottom) vs. smart connected systems (top)

received increasing attention over the last few years [23] as hardware, software, communication, and control units became more advanced and less expensive. As a result, it has created motivations for the firms to integrate a smart connected system to their currently existing platforms.

Figure 1.5 shows the configuration of a company, which employs smart devices and units; however, installing smart devices does not imply a smart connected system. A more important aspect for creating a smart connected system is the ability to connect and communicate either machine to machine (M2M) or machine to human (M2H) with minimal human interventions. Therefore, the infrastructure of smart systems and smart connected systems are different in terms of the communication networks, standards, and protocols [24] as the smart connected system is designed to apply autonomous data-driven real-time decision-making. "Connectivity for everything" is the key feature of a smart connected system [25].

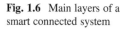

Fig. 1.6 Main layers of a
smart connected system

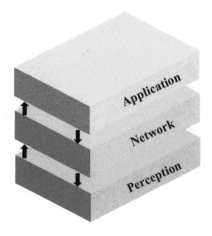

1.2.1   Smart Connected System Main Layers

Smart connected systems include both the cyber and physical systems (CPSs).
Elementary layers of the system start from the physical units. Physical units can be
equipped to be integrated into a smart connected system or can be added using smart
devices. Depending on the selected infrastructure and platform, some modifications
might be needed to make sure the physical units in terms of standards and protocols
are in uniform format. At the same time, the cyber level is more focused on the
communication network including M2M and H2M. These communications com-
prise one to one, one to many, and many to many communications. Figure 1.6
demonstrates the main layers of a smart connected system. The "perception" layer
includes the physical smart devices and considers the interaction between the units
and their integration to the network. Indeed, "things" will be connected to the IoT
network in this layer.

"Network" layer oversees the data transmission over the physical and cyber
systems. What is challenging in a smart connected system including thousands of
smart devices, sensors, actuators, gateways, local edges, central cloud, etc. is the
"navigation" of outputs of these smart devices. Thousands of smart devices in a
smart connected system are continually measuring, collecting, analyzing, and transferring the data streams. The size of the generated data is enormous and might not be
easy to be handled close to real time. This layer determines how smart devices
frequently collect, store, and transfer the data streams from each node of the system.

A timestamp is a time at which a data has been recorded by a computer, and there
should be a slight difference between the time of the event and recorded time. All the
timestamps must be collected in a log file, and the closer the timestamps to the event
time, the more accurate results of the analysis [26]. ISO 8601 standardizes the
representation of dates and times [27]. The data format, which consists of datatype
and the unit, should be determined based on the standards since the human and
machine should easily be able to read the data. Furthermore, communication format

should also be selected based on the standard for Wi-Fi or physical hardware devices. Therefore, both sensors and network should follow the standard and have the same language.

In addition to the time stamping, data routing optimization is performed in the networking layer. The optimum routes for data stream transformation will be determined with respect to several constraints such as bandwidth, security, overhead delay, processing, and transferring cost. Hubs, switching, gateways, cloud, edge, and fog computing are also considered at this level. Furthermore, determining the best location for performing the analytics and making the decision about which part of the data should be transferred to the higher levels are among the crucial decisions of this layer.

"Application layer" is more related to the business and services. This layer delivers the services based on the processed real-time data analytics to provide the required service for the operation. For instance, customers will be able to track their electricity usage, current and expected energy generation, the amount of electricity supplied by the grid, applications usage, etc.

1.3   Dataflow in the Wisdom Pyramid

There are various types of smart devices in the range of small wearable accessories to large machines. Smart devices include a small smart chip, which gives them the ability to be controlled remotely. These small embedded chips are responsible to measure and collect one or more predefined variables. Some of the smart sensors are embedded in a data acquisition unit which is able to store a limited amount of data, transfer the data to higher levels, perform basic data analytics in the local level, and take some preliminary actions.

Dataflow in an IoT system starts from the collected raw data using the smart devices. The types and numbers of smart devices can significantly affect the performance and speed of the system. The measurable variables are important in the network and should be selected beforehand. Based on the selected attributes, smart sensors should be installed or activated on the most important nodes of the system. Some of the units and equipment are already smart using embedded smart devices, and they just need to be integrated into the system in standard structures which have been selected for the network communication.

Figure 1.7 shows the dataflow starting from smart devices to the higher level of a smart connected system. It also connects the dataflow diagram to Fig. 1.1, which are the main elements of a smart connected system. One of the objectives of a smart connected system is monitoring the system as close as possible to the real time which happens in the machine and process data elements in an IoT system. It should be considered that usually raw data do not reveal any helpful information about the status of the system. Analytics should be performed in the control center and data analytics elements to reveal the information, pattern, abnormalities, etc. Key performance indicators (KPIs) are useful to extract beneficial information out of collected

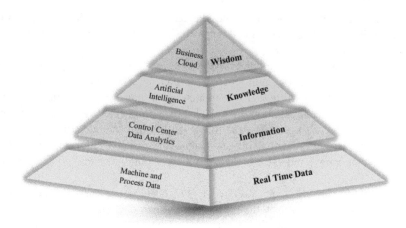

Fig. 1.7 Dataflow using smart devices

Fig. 1.8 Factors which affect the gained knowledge in IoT networks

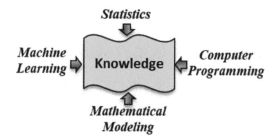

raw data [28]. Efficiency, availability, reliability, and effectiveness are a few examples of the general KPIs which can be applied to most of the systems. In the next step, calculated analytics should be analyzed through AI algorithms to reach knowledge about the system. For instance, using the real-time monitoring techniques, the next potential failure should be predicted using the knowledge of the system based on the trained algorithms. As Fig. 1.8 shows, statistical principles, ML algorithms, mathematical modeling, and computer programming highly affect the quality of the gained knowledge.

In the last layer, business enterprise and cloud computing can transform the gained knowledge into wisdom. Reaching this layer requires the highest level of automation in a smart connected system. Using the output of the AI principles, the system should reach a point which has the acceptable wisdom to take optimal actions. When a smart connected system reaches a stable wisdom point, it should be smart enough to perform most of the tasks automatically.

These actions can be taken for descriptive, diagnostic, predictive, and prescriptive purposes. Descriptive actions reveal and monitor the status of the system before any abnormality in the system. Diagnostic activities modify the performance of the system by taking the necessary actions to prevent any condition which might affect the performance or lifetime of the assets. Predictive actions try to catch the random

Fig. 1.9 The big picture of
a connected smart system
and its components

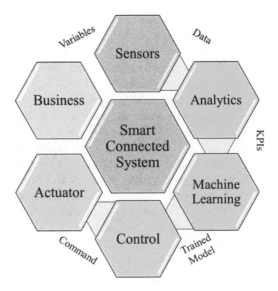

failures in advance in order to be flexible enough to prevent any breakdown in the system. Prescriptive activities are related to the required actions after the failure happened.

1.3.1   Smart Connected Systems Components and Technologies

Figure 1.9 depicts a big picture of a smart connected system component for the real-world applications. At this level, the components can be categorized as sensors, actuators, and control center as well as the communication network equipment.

- A sensor is a transducer that converts a physical stimulus from one form into a more useful form to measure the stimulus. Its purpose is to capture data on events or changes in the surrounding environment and send the information to control centers.
- An actuator is a hardware device that converts a controller command signal into a change in a physical parameter and is responsible for moving and controlling a mechanism or system.
- A data aggregator is a web service which collects the data in an efficient manner with respect to the network lifetime, data accuracy, data latency, and data security. The main goal of sensor data aggregator is to integrate sensors data from data clouds of various manufacturers.
- Gateway is a link between computer programs, aggregators, allowing them to share information and bypass certain protocols.

- Data analytics is performed at several stages in the system. The first step can be in the sensor to convert the raw data to information to be transferred via a digital communication. The next level is on-premise at machine level in a processor such as industrial PC or local data center. The ultimate analysis can be performed at the cloud level with access to several functions such as powerful computation engines and large data storage capabilities.
- Artificial intelligence and machine learning can also be applied at two main levels, on-premise computation engine or cloud. Data analytics and artificial intelligence at machine level can enable fast response for controls and safety, but they require a larger upfront cost for system hardware and software setup.
- Control system operates actuators and devices throughout the system based on data from sensors, on-premise analytics, and cloud analytics. Optimized and adaptive control can be implemented based on real-time and post-processed data.
- Business layer is a crucial level that can include a manufacturing execution system (MES), enterprise resource planning (ERP), or any business intelligence function. It makes the ultimate decision on process, throughput, and operation of the system.

In a smart connected system, the machines at the floor are connected in a bidirectional data loop through data networks to the cloud and business layer. As L. Lee and K. Lee [29] stated in their research article, five major communication network technologies that enable an IoT network to interconnect are as follow:

1. Radio Frequency Identification (RFID)

   - Performs the identification of the units, materials, and equipment using radio frequency wave through the tag and reader systems.

2. Wireless sensor networks (WSN)

   - Measures autonomous variables using smart devices to monitor the status of the units or environmental conditions. WSN should be compatible with RFID platforms [30].

3. Middleware

   - Middleware is a software layer between software applications to make communications easier by hiding the nonrelevant details of different technologies.

4. Cloud computing

   - Indeed, the cloud is a model for on-demand access to a variety of resources, data analytics, stored data, etc. [31]

5. IoT application

   - IoT application allows M2M and H2M interactions through a reliable and secure platform.

1.4   Selection of Smart Devices

What made IoT very attractive for both industry and academia is mostly its focus on the "things" rather than the Internet. Although without the Internet connection none of the communications are possible, the changing nature of the "things" is more radical. Therefore, a systematic approach is necessary to select the most applicable sensors for a specific application among the pool of manufacturers. The most important part of the sensor selection is installing and developing the sensors which exactly meet the predefined requirements. Based on the Oxford English Dictionary "a device which detects or measures some condition or property and records, indicates, or otherwise responds to the information received" is called sensor.

There are wide ranges of choices for sensor selection with a high competition between the manufacturing companies. Developments in chip manufacturing technology lead to smaller, smarter, cheaper, and easier to use sensors. Sensors are also different in their qualities and the amount of noise or variance in their variable measurements. Therefore, sensor selection is an influential step in making a successful smart connected system. Before starting the sensor selection process, it is required to exactly clarify the task of the sensors and specify the characteristics which are going to be measured. Sensors can be categorized based on the phenomenon they are supposed to measure. Based on the application, sensors can be single or multifunctional, and each sensor should be trained for each function before it can be used.

Sensors are also different in their physical characteristics. It is important to find a sensor that matches the requirements of the specific system. Table 1.1 summarizes the main physical characteristics of a sensor.

For instance, the role of the smart devices for "energy" sector is vital. During the last decades, high penetration of the distributed generators (DGs) amplifies the role of digital communication, monitoring, and controls in a smart micro-grid. Utilizing the best controllers and smart devices can significantly enhance the reliability, sustainability, efficiency, and security which are among the most challenging

Table 1.1 Summary of the main characteristics of a sensor [32]

Range	Maximum minus minimum measured stimulus
Resolution	Smallest measurable increment
Sensing frequency	Maximum frequency of the stimulus which can be detected
Accuracy	Error of measurement
Size	Leading dimension or mass of the sensor
Operational environment	Operating temperature and environmental conditions
Reliability	Service life or number of cycles of operation
Drift	Long-term stability
Cost	Purchase cost of the sensor

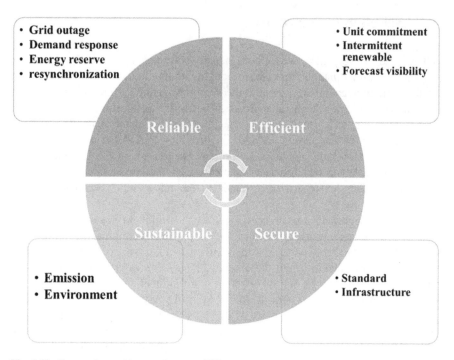

Fig. 1.10 Smart micro-grid control system [33]

concerns of a smart micro-grid [32]. Figure 1.10 depicts the variables that should be measured to reach four objectives of the system. Online metering of the loads generated and stored energy is a key feature to have the dynamic optimization of the system operations. All the decisions regarding the objectives and variables of the system have a direct effect on the sensor selection process.

Based on the infrastructure, size, and target of the system, various characteristics should be considered for the sensors. Defining all these requirements is vitally important since these characteristics can significantly affect future analysis. Furthermore, correct selection of each of the variables as well as the associated sensors to measure them can postpone the system upgrade which has more economic benefits for the system.

To summarize, the answer to the following questions before making any decision for selecting each of the sensors of the system can affect the future performance of the smart devices integrated into the smart connected systems.

1. What are the main "applications"?
2. What are the "characteristics" which should be measured?
3. What are the "ranges" (maximum minus minimum measured stimulus)?
4. What are the "resolutions" (smallest measurable increment)?
5. What are the "accuracies" (error of measurement)?

6. What are the "sensing frequencies" (maximum frequency which can be detected)?
7. What are the "size limits"?
8. What are the "environmental and operating conditions"?
9. What is the desired "reliability" (service life or number of cycles of operation)?
10. What is the "drift" (long-term stability)?
11. What are the "costs" for purchasing the sensors?
12. What is the preference between "single or multifunctional" sensors?
13. What is the desired "timestamp"?
14. What are the "platform modules" (low, medium, or high channel)?
15. What is the "connection interface" (stand-alone or PC connected)?
16. What are the "datatypes" and "unites"?
17. What is the desired "connection" (Wi-Fi or hardware)?
18. What are the most important features of the "software"?
19. What is the "output signal" preference (analog or digital)?

1.5 Centralized and Decentralized IoT Structures

As mentioned before, units are at the base layer of an IoT structure. Next level would be making all the units smart and compatible with the selected platform and standards. Fog and cloud computations are related to the data analytics computation at the decentralized scheme, while the cloud is the center of the computation and data storage. There is an upcoming shift in the current IoT in terms of the structure. New proposed decentralized structures are pushing the data analytics and computations as close as possible toward the end customers [20]. The main advantage of the decentralized computations could be minimizing the overhead delays and monitoring the process as close as possible to real time. In this architecture, the useful part of the collected data should be transferred to a higher level in terms of information, KPIs, and data analytics.

The data stream routing problem is still one of the challenges for researchers. A system should perform the optimization constantly to find out the optimized routes for transferring the data stream with respect to the transferring cost, processing cost, overhead delay, bandwidth, and security. The concepts of the centralized and decentralized computations will be discussed in detail in the following chapters of this book.

Figures 1.11 and 1.12 are presenting a schematic overview of the smart connected system. As they depict, decentralized computations at the local device level (Fog), decentralized computation at series of devices at the system level (Edge), historical data, real-time dataflow, and centralized cloud center are the main parts of a smart connected system with a crucial role for each of the levels.

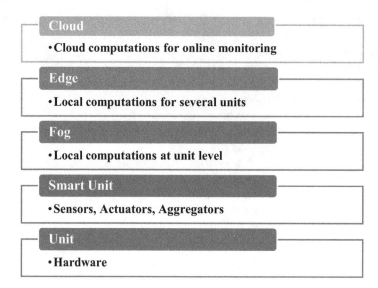

Fig. 1.11 Big picture of a connected smart system architecture

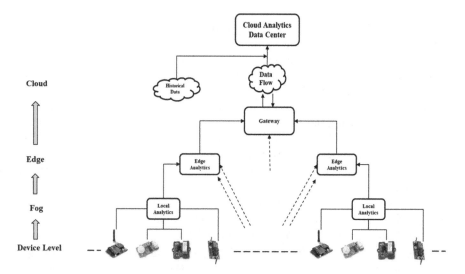

Fig. 1.12 Schematic view of a connected smart system

1.6 Business Models' Perspective

Applications of the IoT are very vast and not limited to the production and manufacturing phases. Kevin Ashton first used the term IoT in 1999 in the supply chain management (SCM) context [4]. Internet of Things (IoT) governs every step of the value chain starting from suppliers, production, services, and guarantee.

Established value chain activities can bring more opportunities and novel business conceptions at the business level to be more successful among the pool of competitors [34]. Kiel (2017) and Brettel et al. (2014) stated that the business models (BM) have got less attention between other areas of the IoT, while there are a lot of challenges and opportunities for the researchers [35]. As stated by several researchers [36–39], the lack of one established business model (BM) is currently challenging.

On the word of Osterwalder et al. (2005), the business model (BM) is a "conceptual tool that contains a set of elements and their relationships and allows expressing the business logic of a specific firm. It is a description of the value a company offers to one or several segments of customers and of the architecture of the firm and its network of partners for creating, marketing, and delivering this value and relationship capital, to generate profitable and sustainable revenue streams" [40]. Weill and Vitale (2001) [41] defined the business models (BM) as the "description of the roles and relationships among a firm's consumers, customers, allies, and suppliers that identify the major flows of product, information, and money and the major benefits to participants." Based on other descriptions, the business models (BM) is the core logic for an organization to create value for the customers which the offered value is better than its competitors while bringing economic benefits for the organization [42, 43]. IoT applications, connectivity, and flow of information between different supply chain partners can be studied and used to create more successful business models [44] .

1.7 Concluding Remarks

Although IoT networks are beginning to appear in our daily life routines and have a huge impact on the quality of our lives, Industrial Internet of Things (IIoT) has gotten more attention since firms without IoT networks will soon be demolished by more competitive IoT-based companies. Firms can use a combination of smart devices with the ability to measure a predetermined variable as well as the capability to exchange data with other machines or human, to create the infrastructure of an IoT network. The data gathered in this network is then used as the raw input to feed the developed algorithms in a way to make the system as smart as possible and to be able to make more optimal and reliable decisions with the minimum amount of human intervention. However, optimal large-scale decisions require the partners in a supply chain to share their IoT networks and create a more intelligent business model.

Several decisions should be made before developing an IoT network to ensure a successful infrastructure. Firms need to decide on the variables about the system status that should be measured, the sensors and devices that have the required quality of measurement and communication abilities, the technologies that are required for data exchange and interactions, and also the decision-making algorithms. Supply chains can benefit a lot from an IoT network if they create IoT structures based on the steps mentioned in this chapter.

References

1. J. Gubbi, R. Buyya, S. Marusic, M. Palaniswami, Internet of Things (IoT): A vision, architectural elements, and future directions. Futur. Gener. Comput. Syst. **29**(7), 1645–1660 (2013)
2. I. Lee, K. Lee, The Internet of Things (IoT): Applications, investments, and challenges for enterprises. Bus. Horiz. **58**(4), 431–440 (2015)
3. M. Hartmann, B. Halecker, Management of innovation in the industrial internet of things. In: Proceedings of the 26th International Society for Professional Innovation Management Conference (ISPIM), Budapest, pp. 1–17 2015.
4. K. Ashton, That 'internet of things' thing. RFID J. **22**(7), 97–114 (2009)
5. D. Miorandi, S. Sicari, F. De Pellegrini, I. Chlamtac, Internet of things: Vision, applications and research challenges. Ad Hoc Netw. **10**(7), 1497–1516 (2012)
6. P.K. Kannan, Digital marketing: A framework, review and research agenda. Int. J. Res. Mark. **34**(1), 22–45 (2017)
7. C. Falkenreck, R. Wagner, The Internet of Things – Chance and challenge in industrial business relationships. Ind. Mark. Manag. **66**, 181–195 (2017)
8. W. Bauer, S. Schlund, D. Marrenbach, O. Ganschar, *Industrie 4.0-Volkswirtschaftliches Potenzial for Deutschland, Studie des Bundesverband Informationswirtschaft, Telekommunikation und neue Medien e* (V.(BITKOM), Berlin)
9. P. De Leusse, P. Periorellis, T. Dimitrakos, S.K. Nair, *Self Managed Security Cell, a security model for the Internet of Things and Services*, First International Conference on Advances in Future Internet. IEEE, pp. 47–52 (2009)
10. D. Kiel, C. Arnold, K. Voigt, The influence of the Industrial Internet of Things on business models of established manufacturing companies – A business level perspective. Technovation **68**, 4–19 (2010)
11. O. Mazhelis, E. Luoma, and H. Warma. *Defining an internet-of-things ecosystem.* Internet of Things, Smart Spaces, and Next Generation Networking. Springer, Berlin, Heidelberg, 1–14 (2012)
12. M. Iansiti, R. Levien, *The keystone advantage: what the new dynamics of business ecosystems mean for strategy, innovation, and sustainability* (Harvard Business School Press, Boston, 2004)
13. M.E. Porter, J.E. Heppelmann, How smart, connected products are transforming competition. Harv. Bus. Rev. **92**(11), 64–88 (2014)
14. L. Abrahams, Regulatory imperatives for the future of SADC's "digital complexity ecosystem". Afr. J. Inf. Commun. Technol. **20**, 1–29 (2017)
15. F. Civerchia, S. Bocchino, C. Salvadori, E. Rossi, L. Maggiani, M. Petracca, Industrial Internet of Things monitoring solution for advanced predictive maintenance applications. J. Ind. Inf. Integr. **7**, 4–12 (2017)
16. Gartner, Gartner says the Internet of Things will transform the data center. Retrieved from http://www.gartner.com/newsroom/id/2684616, March 19
17. V. Krotov, The Internet of Things and new business opportunities. Bus. Horiz. **60**(6), 831–841 (2017)
18. J. Buckley (ed.), *The Internet of Things: From RFID to the Next-Generation Pervasive Networked Systems* (Auerbach Publications, New York, 2006)
19. M.A. Khan, K. Salah, IoT security: Review, blockchain solutions, and open challenges. Futur. Gener. Comput. Syst **82**, 395 (2018)
20. L. Atzori, A. Iera, G. Morabito, The Internet of Things: A survey. Comput. Netw. **54**(15), 2787–2805 (2010)
21. Statista, Internet of Things – number of connected devices worldwide 2015–2025
22. Statista and Smart home – Statistics & Facts, Projected Internet of Things services spending worldwide from 2014 to 2017 (in billion U.S. dollars)
23. F. Tao, Y. Wang, Y. Zuo, H. Yang, M. Zhang, Internet of Things in product life-cycle energy management. J. Ind. Inf. Integr. **1**, 26–39 (2016)

24. M.E. Porter, V.E. Millar, How information gives you competitive advantage. Harv. Bus. Rev. **63**, 149–160 (1985)
25. N.V. Wuenderlich, K. Heinonen, A.L. Ostrom, L. Patricio, R. Sousa, C. Voss, J.G. Lemmink, "Futurizing" smart service: Implications for service researchers and managers. J. Serv. Mark. **29** (6/7), 442–447 (2015)
26. T. Saarikko, U.H. Westergren, T. Blomquist, The Internet of Things: Are you ready for what's coming? Bus. Horiz. **60**(5), 667–676 (2017)
27. K. Chen, P. Zhuang, Disruption management for a dominant retailer with constant demand-stimulating service cost. Comput. Ind. Eng. **61**(4), 936–946 (2011)
28. D. Setijono, J.J. Dahlgaard, Customer value as a key performance indicator (KPI) and a key improvement indicator (KII). Meas. Bus. Excell. **11**(2), 44–61 (2007)
29. P.A. Bernstein, E. Newcomer, *Principles of Transaction Processing* (Morgan Kaufmann Publishers, Burlington, 2009)
30. Anonymous, ISO. 2004-12-01. Retrieved 2010-03-07. 3.5 Expansion . . . By mutual agreement of the partners in information interchange, it is permitted to expand the component identifying the calendar year, which is otherwise limited to four digits. This enables reference
31. Siemens, Microgrids: The road to local energy independence
32. M.H. Balali, et al., *Development of an economical model for a hybrid system of grid, PV and Energy Storage Systems. 2015 International Conference on Renewable Energy Research and Applications (ICRERA)*, (IEEE, 2015)
33. J. Shieh, J.E. Huber, N.A. Fleck, M.F. Ashby, The selection of sensors. Prog. Mater. Sci., 461–504 (2001)
34. G. Fersi, A distributed and flexible architecture for internet of things. Procedia Comput. Sci. **73**, 130–137 (2015)
35. M. Iansiti, K.R. Lakhani, Digital ubiquity:: How connections, sensors, and data are revolution-izing business. Harv. Bus. Rev. **92**(11), 19 (2014)
36. M. Brettel, N. Friederichsen, M. Keller, M. Rosenberg, How virtualization, decentralization and network building change the manufacturing landscape: An industry 4.0 perspective. Int. J. Mech. Ind. Sci. Eng. **8**(1), 37–44 (2014)
37. M.W. Johnson, The time has come for business model innovation. Lead. Lead. **2010**(57), 6–10 (2010)
38. C. Zott, R. Amit, The business model: A theoretically anchored robust construct for strategic analysis. Strateg. Organ. **11**(4), 403–411 (2013)
39. R. Casadesus-Masanell, J.E. Ricart, Competitiveness: Business model reconfiguration for innovation and internationalization. Manag. Res. J. Iberoamerican Acad. Manag. **8**(2), 123–149 (2010)
40. G. George, A.J. Bock, The business model in practice and its implications for entrepreneurship research. Entrep. Theory Pract. **35**(1), 83–111 (2009)
41. P. Weill, M. Vitale, *Place to space: Migrating to eBusiness Models* (Harvard Business School Press, Boston, 2001)
42. J.C. Linder, S. Cantrell, *Changing Business Models: Surveying the Landscape* (Accenture Institute for Strategic Change, Cambridge, MA, 2000)
43. A. Afuah, C.L. Tucci, Crowdsourcing as a solution to distant search. Acad. Manag. Rev. **37**(3), 355–375 (2012)
44. A. Osterwalder, Y. Pigneur, C.L. Tucci, Clarifying business models: Origins, present, and future of the concept. Commun. Assoc. Inf. Syst. **16**(1), 1 (2005)

Chapter 2
Industrial Asset Management and Maintenance Policies

2.1 Asset Management (AM)

An asset is an entity with potential or actual value to an organization [1]. An asset can be a tangible or intangible unit which has economic value to the organization and performs tasks, generates cash flow, reduces expenses, and improves sales in some cases [2]. In the real-world application, manufacturers usually provide an expected value for the lifetime as well as an interval which indicates the lower and higher bounds of the lifetime. The provided predetermined lifetime information is based on the standard working conditions of the units which might not be the case in many situations. Most of the time, manufacturers provide an interval estimate of the lifetime with respect to the suggested maintenances and working conditions. It is wothmentioning that failures may occur before or after the time which maintenance dispatches are scheduled. Therefore, it is crusial to perform the maintenance activities in an optimized time window, while the operational and maintenance costs are minimized.

AM is defined as a continuous process of maintaining, upgrading, acquiring, and operating physical assets cost-effectively, based on a constant physical inventory and condition assessment through inspection [4]. El-Akruti et al. (2013) [6] define the AM as "the system that plans and controls the asset-related activities and their relationships to ensure the asset performance that meets the intended competitive strategy of the organization."

In the current competitive world, failure of the critical units might lead to losing a considerable amount of money, reputation, safety, etc. It should be considered that performing the maintenance more often or less than necessary would also cause excessive costs. Figure 2.1 is showing the optimized interval for performing maintenance activities. Some units might be degraded more quickly due to the inspection or more often maintenance. On the other hand, the critical role of some units might force the planners to schedule the maintenance to be more conservatory. Therefore, making the decision for performing or not performing the maintenance at each period can highly affect the planning and operational cost of the system. The

© Springer Nature Switzerland AG 2020
F. Balali et al., *Data Intensive Industrial Asset Management*,
https://doi.org/10.1007/978-3-030-35930-0_2

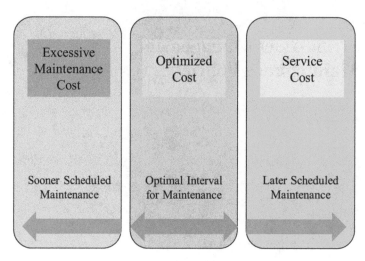

Fig. 2.1 Optimized interval for performing the maintenance

Fig. 2.2 Asset management
(AM) main focuses

optimum schedule is being developed based on the real-time monitoring of the assets due to the dynamic nature of the system. Although the data are collected in real-time basis, algorithms might be run either online or offline based on the main purpose of the AM strategies. It should be considered that the historical data can also highly affect the outputs of the algorithms.

Each firm might have a unique AM approach to manage and maintain its assets align with the one general goal, which is the organization's success. Success can be measured in terms of revenue, cost, lead time, etc. [3]. Indeed, AM balances the costs, opportunities, and risks against the desired performance of assets [5] as it is depicted in Fig. 2.2.

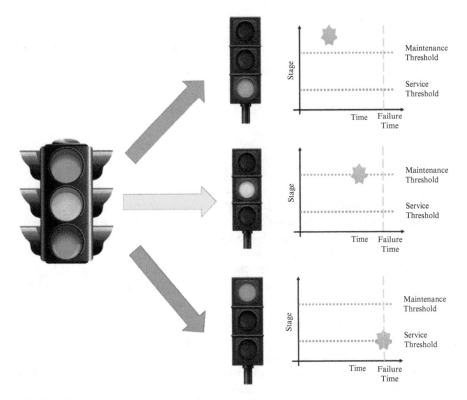

Fig. 2.3 Maintenance of various scenarios

Literature categorizes asset management activities in three main different ways.

1. AM-related activities can be categorized into two main groups as asset life-cycle and supporting activities. Research, design, engineering, development, acquisition, installation, maintenance, replacement, and disposal are examples of the asset life-cycle activities. Examples of the supporting activities are procurement, technical supports, information technology (IT), information systems (IS), finance, accounting, inventory handling, and safety.
2. AM activities can also be sectioned into two groups of maintenance and service activities relating to the time of the action. There is always a threshold value for performing the maintenance as well as service activities. Service refers to the cases which the unit has already failed, which probably is out of order. Figure 2.3 depicts the time spans of these two categories. Green light means that assets are working normally, and no action is required at this time. Yellow light is an alert that the asset might stop working soon; therefore, the problem should be detected and dealt with as soon as possible. Failure happens if the maintenance crew cannot restore the system to the green light during this time. When the red light appears, the breakdown has happened, and it is time to request for the service as soon as possible with respect to the priority of the failed unit. These steps might

Fig. 2.4 Asset management
activities

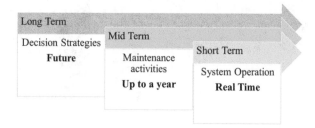

not be real physical light. As an example, lights can be defined as the condition of
the asset in condition-based maintenance (CBM) techniques.
3. Figure 2.4 categorizes the traditional AM activities based on the implementation
   time. Short-time activities include mostly the operation decisions which must be
   made in real time up to weeks. Mid-term activities typically contain decisions
   regarding inspection, replacement, retirement, and maintaining the assets. This
   type of decision usually takes up to a few months. The long-term activities are
   related to the strategical planning and development for the future.

Traditionally, AM activities usually did not consider the status of the units in real-
world applications. However, the working conditions of an asset can highly affect
the performance of the asset in terms of quality. At the same time, the lifetime of an
asset dynamically changes due to the variations in its conditions. Therefore, all types
of assets should be managed and maintained based on an optimal and predetermined
approach to maximize the performance of the asset in terms of efficiency, availability, reliability, maintainability, etc. Alsyouf (2006) and Pinjala et al. (2006) claimed
that there is a lack of studies on the involvement of maintenance policies to positive
performance business [7, 8].

There are a few techniques such as condition-based maintenance (CBM) that take
into account the offline conditions of the assets which are not in real time. However, in
newer asset management methods with a smartly connected environment, the most
significant parameters on the lifetime and the performance of an asset can be
monitored online and close to real time with acceptable flexibility for taking the
necessary actions. This system has a significant potential to influence all the aspects
of asset′s life-cycle activities from design to disposal. These AM activities focus on
controlling the life-cycle conditions of the assets. As Fig. 2.5 shows, El-Akruti et al.
(2013) [9] represented an entire framework of smart asset management system
activities, relationships, and mechanisms. Each step can be further defined as follows:

1. Decision-making: as it was discussed in the first chapter, the measurable variables
   and KPIs should be selected in this step.
2. Plan: at this point, target values should be determined for the KPIs. The smooth
   flow of information and compatibility between the expected life cycle and the
   supporting activities should be ensured.

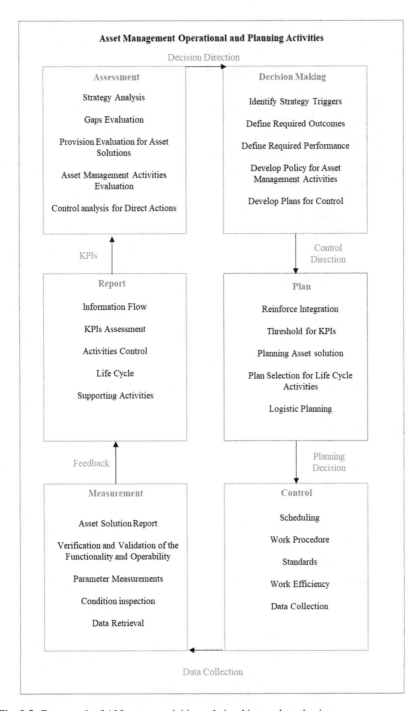

Fig. 2.5 Framework of AM system activities, relationships, and mechanism

3. Control: in this step, work efficiency and effectiveness should be calculated, and all the supporting tasks should be performed to enhance the overall performance of the units. The condition of the assets should be carefully inspected by ensuring that the assets are working based on the predefined standards.
4. Measurement: the data will be collected at certain time stamps. Data retrieval, collection, storage, processing, and defining standard reporting format are among the initial steps of the AM activities.
5. Report: KPIs should be precisely calculated, to transform the collected raw data to useful information which represent the assets' status. In this step, AM activities mostly include maintenance policies by analyzing the interaction between the life-cycle and supporting activities.
6. Assessment: the outcomes of the report processes are considered as the input of the assessment step. Various AM strategies will be analyzed and evaluated in this step to reach to a justified optimal decision for managing the assets. Calling for maintenance (early), following the maintenance schedule (on time), postponing the maintenance (late), and replacing the asset are a few examples of the decisions which can be made in this step.

Current and future AM strategies are mostly based on the data-driven algorithms which continuously try to take the most optimized decisions. Monitoring the health conditions of the assets plays a vital role for critical systems such as the electrical power system which outage of the main components can lead to a blackout. It should be noted that the risk of the failure events for each asset might be defined uniquely in terms of number of customers affected, restoration time, etc.

To monitor assets' conditions, the analyses start with collecting the data at unit level. It should be noted that more data is not always desirable, since data collection, transmission, verification, filtration, and storage are all cost-intensive activities of the data-driven algorithms. Early stage data analyses should be performed on the raw data to find the optimal data collection periods for the algorithms. Algorithms should be run periodically to find the optimized maintenance dispatches for upcoming periods. Each optimal solution is only valid for that specific period since the dynamic nature of the system might change over the next period. Any of the changes should be reflected through the collected data which feed into the AM algorithms. Generally, the results of the algorithms can be categorized into the following activities:

1. No maintenance is needed: Based on the data-driven algorithms which reflect the real-time status of the system, no maintenance is needed for the current or next few periods. Algorithms should be ideally able to predict the future maintenance decisions, but they might be updated over time.
2. Minor maintenance is needed: This activity more often occurs, and the health score of the system will be fully or partially restored after the maintenance is being performed. For instance, changing the oil of the transformers can restore the transformer performance close to its early stages.
3. Major maintenance: In order to perform major maintenance, the unit might not be available for a few periods. This type of maintenance might not occur frequently during the lifetime of the asset especially with the latest technological

development. Most of the time, the units' health index will not be hundred percent restored. This indicates the permanent degradation of the assets during their lifetime.

4. Replacement: Algorithms might offer to replace the unit due to the high operation and maintenance cost for the upcoming periods. The remaining useful lifetime (RUL) of the assets is one of the variables which highly affects the maintenance decisions. RUL can reach its threshold with respect to the calculated cost. Most of the time, replacing the main components of the system is not occurring frequently. For instance, in a power electrical system, the diesel generators and transformers are expected to work for a couple of decades before reaching to the replacement criteria.

2.2 Maintenance Strategies

Generally, organizations consider the maintenance activities as cost evil subordinate to the AM operations [10]. As Bevilacqua and Braglia (2000) [12] reported, 15–70% of the production cost can be caused by maintenance activities. There is always a cost associated with performing the maintenance, as well as a cost of losing an asset due to late maintenance. A considerable portion of this cost usually occurs for the inspection and replacement, operation and idle, as well as failure cost [16]. At the same time, Mobley (2002) [13] claims that one-third of maintenance costs is wasted as a result of unnecessary or improper maintenance activities. Therefore, it is necessary to maintain assets such that the costs are minimized and the useful revenue making the lifetime of the asset is maximized. Recently, the role of maintenance is altering from a "necessary evil" to a "profit contributor" to achieve world-class competitiveness [11].

There are two questions that can be answered by a successful maintenance policy. First, what is the priority of the assets in performing maintenance? The priority of the units that require maintenance is one of the most important parameters for choosing an optimum maintenance strategy for an organization. Some units are vital for a system; therefore, the number of unexpected random failures should be as close to zero as possible. On the other hand, some assets work in parallel, and in case of failure, the system has enough flexibility to successfully continue until the units are back to the working conditions.

Second, when should maintenance actions be performed on each asset? Some believe the outcomes of the maintenance activities are not tangible, and they tend to postpone the maintenance as much as possible [14]. Maintenance costs are another reason that companies tend to delay maintenance. In these cases, the risk of random failure is very high, and these unexpected failures can deteriorate the performance of the system. It is highly important to optimize the time between maintenances to prevent enormous service costs.

In general, it can be concluded that each organization should select an optimal maintenance strategy that pushes the organization toward reaching its goals. The

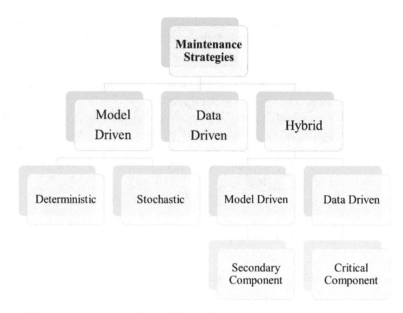

Fig. 2.6 Main categories of the maintenance strategy approach

objective of this optimal maintenance policy is to maximize the overall reliability of the system and availability of the assets with maximum long-term economic benefits [15]. For instance, in a smart manufacturing industry, the optimum maintenance policy can also guarantee the quality of the products as well as the safety of the personnel. Figure 2.6 is presenting the main categories of maintenance strategies.

Historically, industries usually applied the "deterministic" approaches that performed maintenance in fixed intervals or based on the performance threshold which were applied as needed [17]. For instance, the engine oil was replaced every 4 months regardless of how the asset was deployed. As J. Endrenyi et al. (2001) [17] stated, fixed time interval maintenance policies are among the oldest approaches which are easy to employ but not very efficient since maintenance activities directly affect the reliability of the system. If the maintenance activities are performed too rare or too often, the overhead cost will be increased due to more unexpected failures and more inspections and monitoring cost, respectively.

Various factors should be taken into account to reach an optimal maintenance policy. Some of these factors include usage, age, and unexpected failures under different working conditions. Failure can transpire suddenly such as the unexpected change in voltage or can occur over time because of wear, friction, fatigue, erosion, and usage. Therefore, two main categorizes of the failure are random and age-related failures. What is crucial for an organization is the ability to control for these variables, predict any type of failure, and minimize the frequency and duration of the unexpected failures. The ability to predict the failure events is one of the vital factors in determining optimal maintenance dispatch. Failure prediction should be executed with respect to the different parameters such as cost, safety, downtime, etc. If a system can forecast an upcoming failure, it usually has the flexibility to replace or repair the asset with minimum cost and downtime.

Therefore, newer "stochastic" approaches such as "reliability-centered maintenance (RCM)" take the traditional methods and add mathematical modeling techniques. These methods quantify the components' deterioration and enhancements by considering the effects of maintenance activities on system reliability. At the same time, "condition-based strategies" (CBS) are another example of stochastic approaches that start the inspection and replacement activities anytime that the asset reaches a predefined performance level. As an example, the engine oil should be replaced after 5000 miles regardless of how the asset has been operated in this period.

Deterministic models should be replaced by the stochastic ones due to their capability to describe more realistic processes since deterministic approaches are developed based on a few assumptions, which might not reflect the real-world status of the assets. Stochastic approaches can also solve the optimization problems to minimize the cost while maximizing the reliability. For instance, most of the deterministic analyses are assuming some prior knowledge about the rate of failure (λ). Although the results of the model-driven approaches could be applied as a start point of the analysis, it should be considered that they might not always represent the online condition of the assets. On the other hand, data-driven approaches are solely relying on the collected real-time data without any assumption on the performance of the assets. Data-driven analysis usually leads to more optimal solutions since it only relies on the collected data without any prior assumption. Implementation of the data-driven approaches is usually time and cost intensive.

As Wang et al. (2007) [18] presented, "An optimal maintenance strategy mix or 'Hybrid' approach is necessary for increasing availability and reliability levels of production facilities without a great increase in the investment. The selection of maintenance strategies is a typical multiple criteria decision-making (MCDM) problem." Various KPIs can be selected to compare different maintenance strategies. Iterative algorithms are able to derive the most optimal maintenance strategy and the corresponding cost for each approach.

As stated in the literature, maintenance activities can also be categorized into two main groups as corrective and preventive. The corrective maintenance activities occur after the failure, to restore and repair the asset as soon as possible with an acceptable cost and to prevent any loss of generation in products or services. On the other hand, the predictive approaches, which are the main focus of this section, perform the maintenance before the failure point, to keep the equipment in standard working conditions by providing systematic inspection, detection, and prevention of the failure [19]. A system may include thousands of various assets, and it is possible that a single maintenance strategy does not work ideally for all the units since each of them might have a unique behavior over its lifetime, working environment, and safety conditions. Maintenance strategies will be discussed more in detail as follows:

1. *Corrective Maintenance*: Also named "failure-based maintenance" or "breakdown maintenance" since the maintenance is applied after the failure happened

[20]. It can be considered as the origin of the maintenance policies and can be efficiently applied in industries where the profit margin is large [21]. However, more responsive, effective, and reliable strategies are required to be able to compete with other organizations.

2. *Preventive Maintenance*: Also named "time-based maintenance" or "regular time maintenance." Time may refer to calendar time, operating time, or in most cases, the age of the assets. Based on this approach, the maintenance activities should be performed before failures occur. According to the reliability performance of the assets, maintenance crews set periodic maintenances to minimize the frequency and duration of the frequent (expected) and sudden (unexpected) failures.

Since time-based preventive maintenance highly depends on the historical data, the lack of sufficient data may lead to unexpected shutdowns. This is one of the main challenges of this approach since the historical data is not always available or reliable, especially for the newly adopted assets. Another drawback of this strategy is that it usually maintains the assets while a significant amount of life is remaining. This can lead to both unnecessary maintenance and deterioration of the machines if incorrect maintenance is implemented [22].

Condition-based maintenance (CBM) is an example of preventive strategies, which is performed before the failure occurs, and requires a set of variables to be wisely selected and collected frequently. Therefore, this method highly depends on the measured data, mostly collected from smart devices such as sensors. It helps the engineers, managers, maintenance crews, and decision-makers to have a big picture of the system performance over time. Hence, this method also helps to indicate any abnormalities in the assets' working conditions.

Figure 2.7 depicts the functional diagram of the intended CBM system behavior. The first step starts with collecting the data for selected variables in a timely manner and transferring all the digital and analog signals based on the standard language of the system. Diagnostic and prognostics algorithms should analyze the process, based on the decision variable and the measurements. The outcomes of the diagnostic and prognostics algorithms are the main inputs to the decision-making process. After the decisions have been made, the maintenance crew should act and perform the maintenance, replace the parts, or retire and replace the unit. Based on each organization policy, there should be clearly defined threshold values for each of the maintenance activities such as inspection and replacement.

Another example is the predictive maintenance. Some researchers believe that this approach is very similar to CBM [13, 21]. As Wang et al. (2007) stated, "predictive maintenance is able to forecast the temporary trend of performance

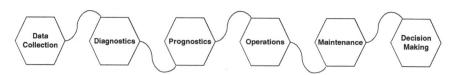

Fig. 2.7 Functional diagram of a CBM system behavior [23]

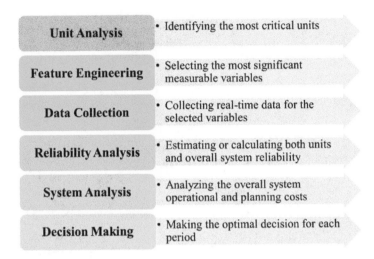

Fig. 2.8 Hybrid maintenance overall framework

degradation and predict faults of machines by analyzing the data collected on the monitored variables. Fault prognostics is a young technique employed by maintenance management, which gives maintenance engineers the possibility to plan maintenance based on the predicted time of future failure and match maintenance activities with production plans, customers' orders and personnel availability" [18].

The new era of the maintenance policies is integrated with the smart devices in order to monitor the assets as close as possible to the real time. The new maintenance policies could be a hybrid approach including different scenarios mostly condition-based maintenance and predictive maintenance. Figure 2.8 summarizes the overall framework to perform the best hybrid maintenance strategy for an organization.

As Fig. 2.9 represents, the current maintenance policy of organizations might be perfect and optimal, or it might be imperfect and need adjustment to reach an optimal approach. In the latter case, two possible cases can happen: (1) another single strategy might work as an optimal approach and (2) a set of strategies need to be combined as a hybrid maintenance policy to reach to an optimal approach. Since there are various types of assets in an organization with different failure, and consequently, different required maintenance activities, it is more likely for an industry to end up with an overall hybrid maintenance policy. An organization might use the same policy for all assets, separate policy for each asset, or a hybrid maintenance policy for all assets.

The only condition is that the selected optimal approach should be able to detect any upcoming failure in the system for each asset, with respect to the priorities, with acceptable flexibility in advance.

Formerly, the main purpose of the maintenance activities was extending the lifetime of the assets. Although it is still one of the most important factors, there are many other goals which have been added to the problem. For instance, in an electrical distribution network, maintenance policies oversee maintaining the

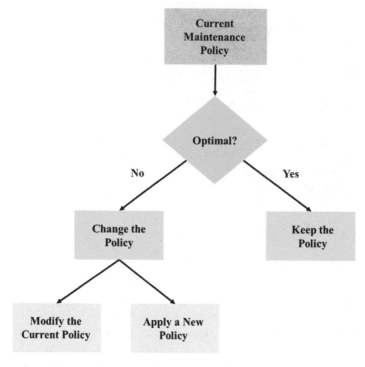

Fig. 2.9 Maintenance policy change diagram

services at an acceptable range. Any type of interruption due to the assets' unexpected failure is not desirable and significantly affects the reliability and availability of the system. Therefore, the newer goal of the maintenance policies can be stated as extending the lifetime of the assets while minimizing unexpected failures. It is expected that an effective maintenance policy should be able to reduce both frequency and duration of the failures.

2.3   Remaining Useful Lifetime (RUL)

As stated in the literature, remaining useful lifetime (RUL) of an asset can be extended using online monitoring techniques while optimizing the maintenance policies and breakdowns. Therefore, it is crucial to study the relationship between the RUL and failure rates of the asset. The lifetime vs. failure rate can be indicated by using a bathtub curve due to the quadratic relationship between the explanatory variables. As W. Roesch (2012) [25] stated, the bathtub curve represented in Fig. 2.10 most likely originated from the studies of human fatality rates. Historically, the curve has been applied to large integrated systems, such as an aircraft or an

Fig. 2.10 A traditional
bathtub curve [25]

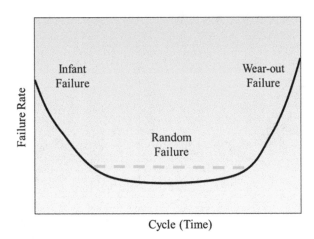

Fig. 2.11 An early or infant
portion of the bathtub curve
[25]

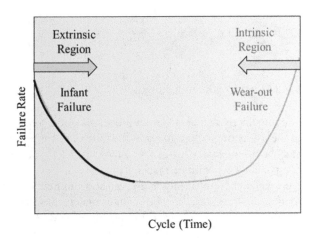

automobile. Because the failure rates are highest during early life and at wear out, the
curve suggests three regions of failure rate.

Scientists divide the lifetime of an asset into these three stages based on the
bathtub curve: (1) early or infant, (2) random, and (3) wear out stage. Figure 2.11 has
often been referred to as the infant region. The rate is assumed to be caused by
defects and constantly decreasing as the population ages. Most early failures are
driven by defects, and typically only a small portion of the population is susceptible.
The study of the early failure distributions is problematic, since the affected population is highly variable and unknown until the entire population is aged into wear
out. As depicted in Fig. 2.12, "wear out is particularly an important region of interest
during the development of new strategies as well as the evaluation of improvements
to existing strategies. Focus on wear out is particularly keen for reliability pro-
fessionals, since wear out mechanisms affect the entire population of the samples,
making the study of failure distributions with small sample sizes possible."

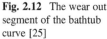
Fig. 2.12 The wear out segment of the bathtub curve [25]

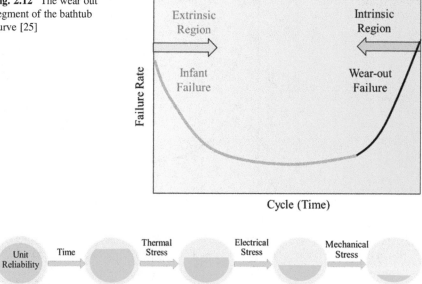

Fig. 2.13 Component reliability modeling

From the overall degradation properties of the assets, it is claimed that the aging of the assets often contributes to the failure, due to electrical, thermal, mechanical, and environmental stresses. X. Zhang and E. Gockenbach (2007) [24] studied component's reliability modeling based on the evaluation of failure statistics and represented their reliability model as shown in Fig. 2.13. Their result shows that the lifetime of the assets can be mathematically modeled as Eq. (2.1). Each of the electrical, mechanical, and thermal stresses, as well as time-related parameters, can be constantly measured using smart devices. Therefore, in a smart connected system, lifetime is not a static number anymore, and it can fluctuate based on the stresses that the asset goes through. Sometimes, the interaction between the assets also plays a significant role since the operating condition of a unit can affect other assets' lifetime and performance. In a smart connected system, the interactions between the units should be considered in addition to monitoring each individual unit.

$$L = L_0 \left(\frac{E}{E_0}\right)^{-(n-bT)} \cdot \left(\frac{M}{M_0}\right)^{-m} \cdot e^{-BT} \qquad T = \frac{1}{\delta_0} - \frac{1}{\delta} \quad [24] \qquad (2.1)$$

where,

L : lifetime

L_0 :   corresponding lifetime.

E : electrical stresses

E_0 : lower limit of electrical stresses

n : voltage-endurance coefficient

b : correct coefficient which takes into account the reaction of materials

T : thermal stresses

M : mechanical stresses

M_0 : lower limit of mechanical stresses

m : mechanical stressendurance coefficient

B : activation energy of thermal degradation reaction

δ_0 : absolute temperature

δ : reference temperature

2.4   IoT-Based Asset Management

A detailed discussion of the Internet of Things (IoT) has been presented in Chap. 1. In this section, the effects of IoT on AM will be deliberated in detail. One of the main advantages of implementing IoT using smart connected devices is the ability to monitor the overall system status close to real time, which delivers outstanding visibility of the components, units, and systems. Figure 2.14 presents the overall view of asset visibility.

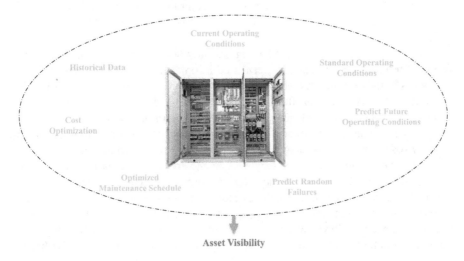

Fig. 2.14 Overall view of asset visibility

- "Current operating conditions" indicate the real-time status of the asset.
- "Standard operating conditions" should be defined based on the standards and manufacturers' suggested policies.
- "Historical data" is considered as the base of the algorithms since the system is relying on the performance of the asset over time. Without considering the historical data, hidden pattern or abnormal cases might be neglected.
- "Predict future operating conditions" based on predictive algorithms. These predictions would be the base for predicting the health indexes for future periods.
- "Predict random failures" as well as the failure due to the aging. There is always a chance that an asset fails while the health indexes are showing the in-control states.
- "Optimize maintenance schedule" based on the real-time data, historical data, current and future working conditions, and the probability of the random failures.
- "Cost optimization" which is the main objective of the algorithm should be done in each period. A cost function can consider the cost of outage and failures as well.

To be competitive, industries need to work on minimizing human intervention and make everything smart in terms of the system component, smart devices, decision-making algorithm, etc. Creating an IoT structure which is smart enough to take the necessary actions automatically as needed is a necessity for companies. In other words, visibility alone is not enough; nowadays, having a successful innovative IoT is a competitive advantage.

Asset management is one of the most important areas that can use the capabilities of IoT for smart decision-making. Smart assets can send the data to a control center. Algorithms can be used to find the patterns in all asset's condition or failure rates, and optimal data-driven decisions can be made in real time to manage and maintain a set of assets in the company with the purpose of minimizing the cost as well as maximizing the reliability of the system which contributes to increasing the revenue. Some researchers call the integration of the maintenance policies and Industry 4.0 as Maintenance 4.0.

Traditional asset management has significantly been altered in a smart environment. The term "traditional" has been used in order to differentiate between AM activities in the presence and absence of a smart connected system or IoT network. Real-time observability of the system components and existence of a smart connected system can significantly change the definition of AM activities.

Formerly, maintenance operators started maintaining and controlling the assets mostly based on the manufacturer suggested plans. After a certain amount of time passed in the lifetime of an asset, they were able to update their approaches using the collected historical data. Sometimes, they performed the maintenance sooner than needed, and some other times, failure occurred before the scheduled maintenance. None of the abovementioned strategies are optimum in terms of life-cycle and maintenance cost. Nowadays, the goal of the asset management strategies is to predict the failure in advance with acceptable flexibility to schedule the maintenance as needed.

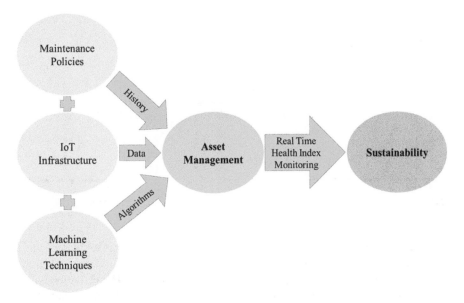

Fig. 2.15 Asset management in a smart connected system

Recently, the definition and application of AM activities have been changed due to the wide application of smart devices and artificial intelligence (AI). As Fig. 2.15 shows, conventional maintenance policies are only one input for the current AM techniques. IoT infrastructures and machine learning (ML) algorithms are other requirements of industry in developing a successful AM strategy. A combination of all these three factors combining with the manufacturer's maintenance plans alongside historical data, collected real-time data, and model-driven algorithms should be applied simultaneously to reach an optimal strategy. Real-time monitoring health index along with the AI can ensure continuous improvement of the AM strategy can aid an organization to have a more sustainable environment in terms of the Real-time monitoring health index along with the AI approaches can ensure continuous improvement of the AM strategy to have a more sustainable environment.

2.5   General Key Performance Indicators

An IoT-based asset management requires a set of KPIs to be defined, monitored, and managed through the time. The main objectives of all AM activities are to enhance the reliability of the system as well as the availability of the assets. It is very important to have a clear definition of all these KPIs to create information out of raw data, address the issue, and formulate a mathematical model. In this section, a few general KPIs will be briefly discussed. It should be considered that the KPIs must be tailored for each application.

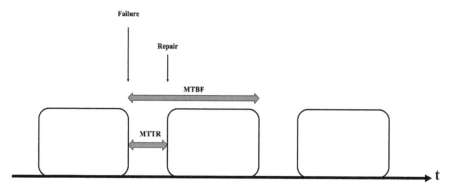

Fig. 2.16 Relation between the failure and repair events with MTBF and MTTR

- *Reliability*: Reliability is the probability that an asset performs in the range of a given specific interval under predefined conditions [26]. Failure rate $\lambda(t)$ shows the total number of failures given in a specific time interval. Therefore, the failure rate is a function of time. It could be constant over time or has an increasing (aging) or decreasing (weed-out) trend.

$$\text{Reliability} = 1 - \text{Probability of Failure}$$

Reliability will be discussed in detail in the next chapters of this book.

- *Availability*: Availability is a measure of the percentage of the time which an asset is operable in a good condition which does not affect the quality [27]. It should be considered that the reliability is

$$\text{Availability} = \frac{\text{Time which an asset is in the working condition}}{\text{Total amount of time}}$$

- *Maintainability*: Maintainability stands for the duration of maintenance. It includes all the needed actions to return an item to the working conditions. Mean time to repair (MTTR) is often used to limit the maximum repair time [28]. Figure 2.16 shows the relation between the failure and repair events with MTBF and MTTR.

  - *Mean time to repair (MTTR)* is a measure which shows the downtime of each component from the time the failure happened to the time that the component is working with the standard conditions [29].
  - *Mean time between failure (MTBF)* states the expected amount of time between failures of an asset and is often reported since it can be interpreted easily [30].

- *Efficiency:* Efficiency measures the productive work output versus the work input.
- *Utilization:* Utilization is the ratio of time spent on productive value-added actions to the total time consumed.
- *Capability*: Capability shows how well the production is performed.

$$\text{Capability} = \text{Efficiency}\,(E) \times \text{Utilization}\,(U)$$

- *Effectiveness*: Effectiveness indicates the overall performance status of the units or systems.

$$\text{Effectiveness} = \text{Availability} \times \text{Reliability} \times \text{Maintainability} \times \text{Capability}$$

2.6   Concluding Remarks

Traditional AM strategies seem not to be efficient since techniques such as fixed time interval maintenances can lead to too often or too rare maintenances. Both of these cases can cause a lot of costs to a company. On the other hand, newer condition or reliability-based methods that set maintenance and service policies based on the historical data and offline status of the system can be more efficient and less costly to the company.

However, none of these methods are as efficient as an IoT-based AM strategy. Using smart devices to report on the online condition of the asset and also online controlling center equipped with trained machine learning techniques can result in an optimal AM strategy for the multi-objective problem of minimizing the cost as well as maximizing the reliability of the asset in its lifetime. An optimal solution to this problem can also extend the lifetime of the asset and increase the quality of the produced products. A smart connected system can similarly make optimal decisions about the priority of the assets since it collects and consumes the data on the status of all the assets in the system.

References

1. N. Bontis, Assessing knowledge assets: A review of the models used to measure intellectual capital. Int. J. Manag. Rev. **3**(1), 41–60 (2001)
2. S.G. Winter, Knowledge and competence as strategic assets. In *The Competitive Challenge: Strategies for Industrial Innovation and Renewal*. D.J. Teece (ed.) (Harper and Row: N.Y.), pp. 165–187 (1987)

3. J.E. Amadi-Echendu, Managing physical assets is a paradigm shift from maintenance
4. K.B. Misra, Maintenance engineering and maintainability: An introduction, in *Handbook of Performability Engineering*, pp. 755–772 (2008)
5. A.H. Tsang, Strategic dimensions of maintenance management. J. Qual. Maint. Eng. **8**(1), 7–39 (2002)
6. K. El-Akruti, R. Dwight, T. Zhang, The strategic role of engineering asset management. Int. J. Prod. Econ. **146**(1), 227–239 (2013)
7. S.K. Pinjala, L. Pintelon, A. Vereecke, An empirical investigation on the relationship between business and maintenance strategies. Int. J. Prod. Econ. **104**(1), 214–229 (2006)
8. I. Alsyouf, Measuring maintenance performance using a balanced scorecard approach. J. Qual. Maint. Eng. **12**(2), 133–149 (2006)
9. K. El-Akruti, R. Dwight, T. Zhang, The strategic role of engineering asset management. Int. J. Prod. Econ. **146**(1), 227–239 (2013)
10. P. Muchiri, L. Pintelon, Performance measurement using overall equipment effectiveness (OEE): Literature review and practical application discussion. Int. J. Prod. Res. **46**(13), 3517–3535 (2008)
11. G. Waeyenbergh, L. Pintelon, A framework for maintenance concept development. Int. J. Prod. Econ. **77**(3), 299–313 (2002)
12. M. Bevilacqua, M. Braglia, The analytic hierarchy process applied to maintenance strategy selection. Reliab. Eng. Syst. Saf. **70**(1), 71–83 (2000)
13. R.K. Mobley, *An introduction to predictive maintenance*, Elsevier, (2002)
14. P. Castka, M.A. Balzarova, C.J. Bamber, J.M. Sharp, How can SMEs effectively implement the CSR agenda? A UK case study perspective. Corp. Soc. Responsib. Environ. Manag. **11**(3), 140–149 (2004)
15. M. Faccio, A. Persona, F. Sgarbossa, G. Zanin, Industrial maintenance policy development: A quantitative framework. Int. J. Prod. Econ. **147**, 85–93 (2014)
16. C.T. Lam, R.H. Yeh, Optimal maintenance-policies for deteriorating systems under various maintenance strategies. IEEE Trans. Reliab. **43**(3), 423–430 (1994)
17. J. Endrenyi, S. Aboresheid, R.N. Allan, G.J. Anders, S. Asgarpoor, R. Billinton, N. Chowdhury, E.N. Dialynas, M. Fipper, R.H. Fletcher, The present status of maintenance strategies and the impact of maintenance on reliability. IEEE Trans. Power Syst **16**(4), 638–646 (2001)
18. L. Wang, J. Chu, J. Wu, Selection of optimum maintenance strategies based on a fuzzy analytic hierarchy process. Int. J. Prod. Econ. **107**(1), 151–163 (2007)
19. H. Wang, A survey of maintenance policies of deteriorating systems. Eur. J. Oper. Res. **139**(3), 469–489 (2002)
20. L. Swanson, Linking maintenance strategies to performance. Int. J. Prod. Econ. **70**(3), 237–244 (2001)
21. R.K. Sharma, D. Kumar, P. Kumar, FLM to select suitable maintenance strategy in process industries using MISO model. J. Qual. Maint. Eng. **11**(4), 359–374 (2005)
22. C.K. Mechefske, Z. Wang, Erratum to "using fuzzy linguistics to select optimum maintenance and condition monitoring strategies". Mech. Syst. Signal Process. **18**(5), 1283 (2004)
23. S.M. Asadzadeh, A. Azadeh, An integrated systemic model for optimization of condition-based maintenance with human error. Reliab. Eng. Syst. Saf. **124**, 117–131 (2014)
24. X. Zhang, E. Gockenbach, Component reliability modeling of distribution systems based on the evaluation of failure statistics. IEEE Trans. Dielectr. Electr. Insul. **14**(5), 1183 (2007)
25. W.J. Roesch, Using a new bathtub curve to correlate quality and reliability. Microelectron. Reliab. **52**(12), 2864–2869 (2012)
26. K. Krippendorff, Reliability, in *Content Analysis; An Introduction to its Methodology*, (Sage Publications, Beverly Hills, 1980)
27. P.D. Coley, J.P. Bryant, F.S. Chapin, Resource availability and plant antiherbivore defense. Science **230**(4728), 895–899 (1985)

28. B.S. Blanchard, D. Verma, E.L. Peterson, J.W. Maintainability, *Maintainability*, (Wiley, Sons, New York, 1995)
29. S.H. Sim, J. Endrenyi, Optimal preventive maintenance with repair. IEEE Trans. Reliab. **37**(1), 92–96 (1988)
30. J.D. Kalbfleisch, R.L. Prentice, *The Statistical Analysis of Failure Time Data*, vol 360 (Wiley, Hoboken, 2011)

Chapter 3
Asset Aging Through Degradation Mechanism

3.1 Asset Degradation

One of the primary concerns of each system is the ability to enhance the reliability of the individual assets and, consequently, the overall system. In general, a reliable asset is able to perform its intended function under the specific working conditions over the predefined cycles, known as the asset lifetime [1]. The intended function of an asset might include various roles. As a result, an asset might be still working while being considered as an unreliable asset. Specific working conditions are usually provided by the manufacturers or can be obtained as the results of laboratory tests [2]. The specific working conditions are usually the nominal conditions which enable the owners to obtain more value of the asset by enhancing its life-cycle characters. Furthermore, an expected asset lifetime is provided by manufacturers or estimated by decision-makers. It should be considered that an asset might face completely a unique condition over its upcoming cycles. On the other hand, an asset might behave entirely different in each application [3]. Therefore, a more realistic measurement should consequently estimate the lifetime of the assets in order to optimize the assets' value by keeping the reliability of the network above a certain level. Traditionally, the lifetime of an asset could be determined using the three main approaches as follows:

(i) Manufacturers' suggestion
(ii) Laboratory results
(iii) Well-defined standards

Obtaining the lifetime data is not always possible due to various reasons. For instance, technological development might lead to zero or a few failures during the test periods. Although an accelerated condition might increase the chance of observing the failure event, it should be considered that some of these tests are destructive which is not desirable especially for the expensive units [4]. It should be noted that these models are deterministic and might not be able to reflect the actual behavior of

© Springer Nature Switzerland AG 2020
F. Balali et al., *Data Intensive Industrial Asset Management*,
https://doi.org/10.1007/978-3-030-35930-0_3

the asset during its lifetime. In addition to that, the same asset might behave totally different in each application. Maintenance policies are scheduled based on the lifetime analysis of each asset. The scheduled maintenance could be categorized as follows based on the time of the scheduled maintenance and needed maintenance:

(i)   Too often, more than actual needs
(ii) Too rare, less than actual needs which leads to failure
(iii) Within an optimized interval which enhances the overall life-cycle cost of the network

The needs for applying more realistic methods became obvious during the last few years. The wide application of the smart devices leads to more connectivity in the network which could significantly support the idea. An optimized methodology should consider the real-time health status of an asset as well as its history in order to obtain optimized system reliability [5]. It should be noted that collecting several measurements might not be able to reveal the most beneficial information. Therefore, the raw measurements of asset characteristics should be mapped to healthy scores, such as degradation value, in order to enhance the ability of the analytics to detect the upcoming failures or unreliable events [6]. As mentioned earlier, the traditional methodologies are not providing an optimum estimate of the reliability since they are not able to reflect the health status of the assets. High penetration of the smart devices such as sensors and actuators offers a high-volume and multidimensional time-series data sets which might enable the analyst to obtain an accurate real-time insight regarding how the assets perform [7]. The degradation models should be able to predict the future condition of the assets based on the behaviors of the health indicators over time. As a result, it can be concluded that the degradation models are the primary point of interest when the analyst needs to conduct the real-time data-driven analysis based on the asset condition rather than the historical lifetime data [8].

Failure definition in degradation-based models is completely different than traditional models. Collected lifetime data for the traditional algorithms are representing the actual physical failure of the asset. In degradation-based analyses, failure events are said to occur since as a point of time which the degradation estimate hits the degradation threshold for the first time [9]. Failure events are usually defined as a point of time when the degradation profile hits the predefined critical limit or threshold. Therefore, predictive algorithms should be able to predict the critical failure time based on the first hitting time models to initiate maintenance before the failure occurs. Figure 3.1 presents the difference between hard physical and soft failure [10]. After the hard failure happens, the unit is not functional anymore, and it might be possible that the whole asset or system will be down due to the failure of the critical units. On the other hand, when the soft failure happens, the unit might be still functional, but the maintenance activities have been initiated and the maintenance crew should be aware of the error caused in that unit. If the necessary actions do not take place right in time, the soft failure would become a hard physical failure. It should be considered that if the fault in this unit is detected, monitored, and solved before the hard failure, the effect of the unit failure on the

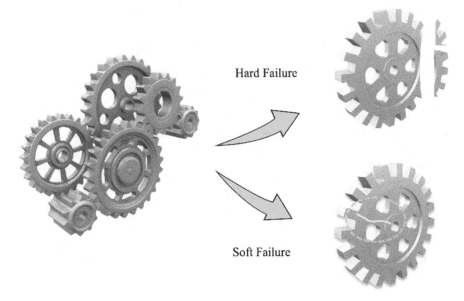

Hard Failure

Soft Failure

Fig. 3.1   Hard physical failure versus soft failure

other assets and overall system would be minimum. Indeed, degradation models are trying to predict the point which the degradation profile reaches to a critical limit which the soft failure happens [11]. For this reason, this point is called the pseudo-life data point since the failure is said to be occurring based on the engineering insight.

3.2   Challenges of Classical Models

High penetration of the smart devices, gateways, communication networks, machine learning (ML) algorithms, predictive models, etc. could highly enhance the performance of the asset management (AM) approaches. When no information exist regarding how an asset behaves in an application, the obtained lifetime data would certainly be the most applicable option [12]. The analyst tried to enhance the performance of the classical models by increasing the number of assets under the study during the testing phase. For instance, it might be possible that analysts observed and recorded the behavior of the hundreds of units until the physical failure happens. The most important challenges of the classical models to estimate the asset condition over time can be summarized as follows:

- Lifetime data cannot be always easily obtained.
- Developed models based on historical lifetime data sets are not able to express the real behavior of the assets.

- Technological developments lead to a few or zero failure data sets as the result of the analysis.
- Some of the laboratory tests and accelerating monitoring actions are expensive.

In degradation-based model, analysts try to predict the condition of the asset over time. It should be noted that there are still some challenges regarding degradation-based models. For instance, the assumptions which the analysts assume regarding the future behavior of the asset might make the prediction inaccurate. If the overall behavior of the degradation process is being observed or known prior, the analysis would be more reliable. For instance, assume that the early readings of transformer conditions reveal a linear relationship between the time and age with the degradation value [13]. If the analysts do not have sufficient insight regarding the future behavior of the degradation profile, they might assume that the behavior remains the same over time, which might not be always true. It should be noted that evolution of degradtion profile may varies in different points of time [14]. Therefore, it can be concluded that predicting the degradation profile can be more mature as the monitoring time passes, the number of same assets under study increases, and the degradation profile reaches its critical limit.

3.3   Prognosis and Health Management (PHM)

PHM is an intelligent condition monitoring approach to:

- Predict the future units' condition
- Predict event which system no longer performs its intended functions
- Estimate time to failure

PHM is a discipline of technologies and methods with the potentials of enhancing reliability estimation that has been revealed due to complexities in environmental and operational usage conditions as well as the effects of maintenance actions [15]. Over the last few decades, several types of research have been conducted in PHM of information and electronics-rich systems as a means to provide advance warnings of failure, enable predictive maintenance, improve system identification, extend system life, and diagnose intermittent failures that can lead to field failure return exhibiting no-fault-found symptoms [16].

Operational availability of information regarding the behavior of the systems has been historically difficult to achieve. The main reason is because of the lack of understanding of the interactions between the various covariates, application environments, and their effect on the system degradation over time. Consequently, there is a pressing need to develop new methods that apply in situ system operational and environmental conditions to detect performance degradation. The most promising discipline of methods with the potential of enhancing the reliability, availability, and maintainability prediction is called PHM [17]. Traditionally, PHM has been implemented using methods that are either model-based or data-driven. The model-based approaches consider the physical processes and interactions between components in the system. The data-driven approaches use statistical pattern recognition and machine learning (ML) to detect changes in parameter data, thereby enabling diagnostic and prognostic measures to be calculated.

Data-driven techniques are utilized to learn from the data and intelligently provide valued information to enhance the decision-making process. They assume that the statistical characteristics of the system remain relatively stable until a faulty condition arises in the system. Anomalies, trends, or patterns can be detected in the collected data by in situ monitoring to determine the health state of a system. The trends are then beneficial to predict the time to failure of the system [18].

Health assessment is carried out in real time using the in situ data and anomaly detection techniques. Knowledge regarding the physical processes in the system and steady-state conditions can help to select the appropriate data analytic techniques. ML is one of the methods to implement anomaly detection techniques, in which the monitored data are compared in real time against a healthy baseline to check for possible anomalies. This is a semi-supervised learning approach wherein data representing all the possible healthy states of the system are assumed to be available a priori. The healthy baseline consists of a collection of data for parameters that represent all the possible variations of the healthy operating states of a system. The baseline data is collected during various combinations of operating states and loading conditions when the system is known to be functioning normally. The baseline can also consist of threshold values based on specifications and standards. It is important that the baseline data should not contain any operational anomalies. The presence of anomalies in the baseline affects the definition of healthy system behavior and hence causes the misclassification of data.

3.4 Degradation Process

Degradation mechanism is an inherent process of a system affected by internal and external factors such as environmental and operational conditions. In simple words, degradation models are trying to capture the accumulated damage over time [19]. During the last few decades, extensive studies have been conducted on the degradation models and their application in the field of reliability analysis, etc. Compared with the traditional lifetime data, degradation estimates might offer more valuable information regarding the asset condition over time. In general, the degradation data might be observed during the laboratory tests, in-service applications, or real-time applications given the normal or accelerated working condition. Degradation data might be collected as the direct measurement (crack length) or measurement of characteristics which directly affect the unit degradation mechanism (vibration or temperature).

It is highly possible that an asset has been degraded due to several mechanisms based on various covariates such as voltage, temperature, vibration, etc. As a result, it is not always possible to present the overall degradation mechanism of an asset

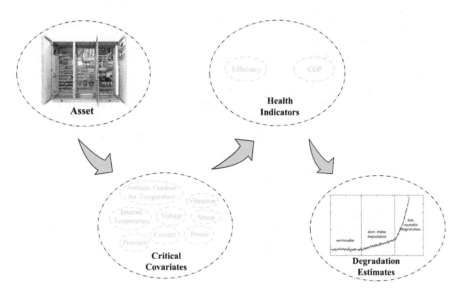

Fig. 3.2 Big picture of an example degradation process

using one profile. An accurate analysis is needed to detect the interaction and dependency of the degradation mechanism along with their effect on failure propagation. It should be noted that constructing an accurate degradation model based on the true degradation process can highly affect the upstream analysis.

There is a growing demand to validate asset reliability relatively quickly with minimal testing. As mentioned before, it might be possible that the output of the analysis includes zero or a few failure events due to technological developments. Therefore, it might not be ideal to assess the reliability based on the traditional approaches. Alternatively, degradation measures will contain beneficial information regarding asset performance and reliability. Degradation measures are applied to estimate the unit condition over time by:

- Preventing an unexpected failure
- Enhancing the service of life, prognosis, and monitoring
- Optimizing the reliability, availability, and risk
- Analyzing the component before the actual failure point

Figure 3.2 depicts an overview of an example degradation process. Consider that the goal is to obtain the degradation measures of an asset over time based on the history of the assets for the most critical internal and external explanatory variables. It has been assumed that the acceptable ranges for the critical variables are known prior. Otherwise, an analysis should be performed in order to select the most critical variables which are affecting the asset aging or degradation mechanism. Any deviation from the acceptable area might be an indication for upcoming failure or unpleasant events. As mentioned before, raw data might not reveal very beneficial information. They can convert to the health indicator statistics in order to extract the

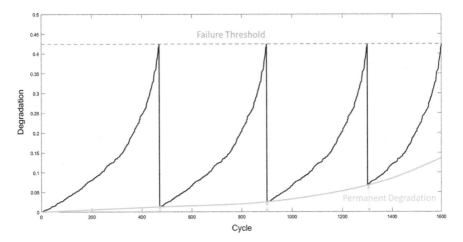

Fig. 3.3 Major types of degradation estimates

most helpful information. As an example, suppose that an electrical transformer is under the monitoring process. Based on initial analyses, a few covariates such as voltage, current, temperature, etc. have been determined as critical covariates which directly affect the degradation measures. In the next step, the raw measurements can be turned into the health indicator statistics such as efficiency. Based on the steady-state conditions, the efficiency of the transformer can be collected and monitored over time. Any abnormal trend in the health indicator estimates might be an indication of forthcoming failure.

As Fig. 3.3 presents, degradation measures can be classified as transitory or permanent degradation. Transitory degradation measures are a part of the degradation process which can be restored by performing the needed maintenances. In some cases, maintenance power is not able to restore the asset to as good as new condition. The difference between the degradation estimates at time zero, when the unit is new, and degradation right after performing the maintenance can be considered as the permanent degradation. Permanent degradation can be used as an indication for replacing the asset over its lifetime.

Degradation models can be categorized as model-driven, data-driven, and hybrid or fusion models. Model-driven algorithms are mostly based on the physics governing the system. Therefore, mathematical models are needed in order to reach a reliable result. Data-driven algorithms are very applicable when the system is complex or the physics of the failure governing on the system is unknown. Data-driven algorithms rely on the data sets which present the system behaviors. ML algorithms are very applicable in order to perform the statistical data analyses and extend outcomes over time. Hybrid or fusion models are based on a combination of model and data-driven algorithms. It means that the mathematical models are able to present the physics of the failure for some parts of the system. Data-driven algorithms are in charge of the complex parts of the system (Fig. 3.4).

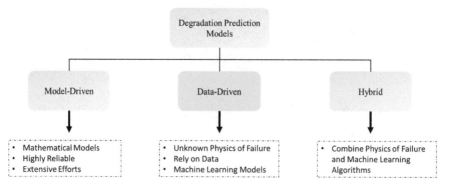

Fig. 3.4 Categories of degradation prediction

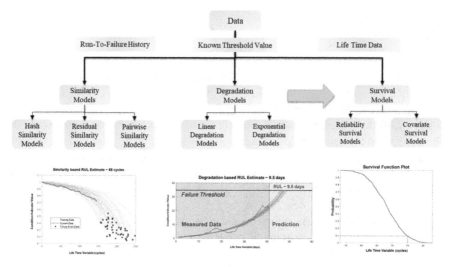

Fig. 3.5 Various categories of the RUL estimation methods

3.5   Remaining Useful Life (RUL)

The remaining useful life (RUL) of a machine is the expected life or usage time remaining before the machine requires repair or replacement. Predicting remaining useful life from system data is a central goal of predictive-maintenance algorithms [20]. The term lifetime or usage time here refers to the life of the machine defined in terms of whatever quantity you use to measure system life. Units of the lifetime can be quantities such as the distance traveled (miles), fuel consumed (gallons), repetition cycles performed, or time since the start of operation (days). Similarly, time evolution can mean the evolution of value with any such quantity. Figure 3.5 depicts various categories of RUL estimation methods [21]. Typically, the RUL of a system is estimated by developing a model that can perform the estimation based on the time evolution or statistical properties of condition indicator values, such as [22]:

- A model that fits the time evolution of a condition indicator and predicts how long it will be before the condition indicator crosses some threshold value indicative of a fault condition.
- A model that compares the time evolution of a condition indicator to measured or simulated time series from systems that ran to failure. Such a model can compute the most likely time to failure of the current system.

3.6 Concluding Remarks

Analysis of the degradation process is one of the lately developed approaches in order to obtain the information regarding the asset condition especially for highly reliable systems, critical assets, and recently developed products. The main purpose of the degradation-based models is to predict the future condition of the asset based on the behavior of the degradation indicators, time, and explanatory variables. Since the degradation-based analysis defines the failure events based on the predefined threshold and critical limits, the failure is said to have occurred as a soft failure. Obtaining the degradation estimates can be considered as one of the challenges of the analysis. It should be noted that each asset might have more than one degradation mechanism. The main purpose of the most recent studies is to develop degradation-based models to predict the critical time for initiating the maintenance actions. The analyses mostly focus on the critical components rather than all components of the systems.

References

1. A.E. Abu-Elanien, M. Salama, Asset management techniques for transformers. Electr. Power Syst. Res. **80**(4), 456–464 (2010)
2. X. Zhang, E. Gockenbach, Asset-management of transformers based on condition monitoring and standard diagnosis. IEEE Electr. Insul. Mag. **24**(4), 26–40 (2008)
3. A. Jahromi, R. Piercy, S. Cress, J. Service, W. Fan, An approach to power transformer asset management using health index. IEEE Electr. Insul. Mag. **25**(2), 20–34 (2009)
4. H. Ge, S. Asgarpoor, Reliability and maintainability improvement of substations with aging infrastructure. IEEE Trans. Power Delivery **27**(4), 1868–1876 (2012)
5. J. Jalbert, R. Gilbert, Y. Denos, P. Gervais, Methanol: A novel approach to power transformer asset management. IEEE Trans. Power Delivery **27**(2), 514–520 (2012)
6. T. Van der Lei, P. Herder, Y. Wijnia, *Asset Management: The State of the Art in Europe from a Life Cycle Perspective* (Springer, Dordrecht, 2012)
7. M. Elcheikh, M.P. Burrow, Uncertainties in forecasting maintenance costs for asset management: Application to an aging canal system. ASCE-ASME J. Risk Uncertain. Eng. Syst. Part A: Civ. Eng **3**(1), 04016014 (2017)

8. A. Abiri-Jahromi, M. Parvania, F. Bouffard, M. Fotuhi-Firuzabad, A two-stage framework for power transformer asset maintenance management—Part II: Validation results. IEEE Trans. Power Syst **28**(2), 1404–1414 (2013)

9. X. Zhang, J. Zhang, E. Gockenbach, Reliability centered asset management for medium-voltage deteriorating electrical equipment based on Germany failure statistics. IEEE Trans. Power Syst **24**(2), 721–728 (2009)

10. J. Tang, T. Su, Estimating failure time distribution and its parameters based on intermediate data from a Wiener degradation model. Nav. Res. Logist. **55**(3), 265–276 (2008)

11. X. Wang, P. Jiang, B. Guo, Z. Cheng, Real-time reliability evaluation with a general Wiener process-based degradation model. Qual. Reliab. Eng. Int. **30**(2), 205–220 (2014)

12. R. Narayanrao, M.M. Joglekar, S. Inguva, A phenomenological degradation model for cyclic aging of lithium ion cell materials. J. Electrochem. Soc. **160**(1), A137 (2013)

13. K. Häusler, U. Zeimer, B. Sumpf, G. Erbert, G. Tränkle, Degradation model analysis of laser diodes. J. Mater. Sci. Mater. Electron. **19**(1), 160–164 (2008)

14. D. Pan, J. Liu, J. Cao, Remaining useful life estimation using an inverse Gaussian degradation model. Neurocomputing **185**, 64–72 (2016)

15. M. Pecht, Prognostics and health management of electronics. *Encyclopedia of Structural Health Monitoring* (2009)

16. K. Goebel, B. Saha, A. Saxena, J.R. Celaya, J.P. Christophersen, Prognostics in battery health management. IEEE Instrum. Meas. Mag. **11**(4), 33–40 (2008)

17. K.L. Tsui, N. Chen, Q. Zhou, Y. Hai, W. Wang, Prognostics and health management: A review on data driven approaches. Math. Probl. Eng. **2015**, 793161 (2014)

18. D. Kwon, M.R. Hodkiewicz, J. Fan, T. Shibutani, M.G. Pecht, IoT-based prognostics and systems health management for industrial applications. IEEE Access **4**, 3659–3670 (2016)

19. M. Giorgio, M. Guida, G. Pulcini, An age-and state-dependent Markov model for degradation processes. IIE Trans. **43**(9), 621–632 (2011)

20. R. Khelif, B. Chebel-Morello, S. Malinowski, E. Laajili, F. Fnaiech, N. Zerhouni, Direct remaining useful life estimation based on support vector regression. IEEE Trans. Ind. Electron. **64**(3), 2276–2285 (2017)

21. X. Si, W. Wang, C. Hu, D. Zhou, Remaining useful life estimation–a review on the statistical data driven approaches. Eur. J. Oper. Res. **213**(1), 1–14 (2011)

22. N. Chen, K.L. Tsui, Condition monitoring and remaining useful life prediction using degradation signals: Revisited. IIE Trans. **45**(9), 939–952 (2012)

Chapter 4
Predictive Degradation Models

4.1 Degradation Models

Asset degradation or deterioration refers to a decrease in the healthy condition estimates of an asset over time. For some applications, the estimates of the health conditions can be achieved by directly measuring a characteristic such as the length of a crack. For some other applications, characteristics representing the system status should be mapped into the degradation estimates. It might not be easily achievable for real-world applications [1]. Analysts need to have deep insight regarding how the system under study is working. The possible potential and most commonly occurring failure events should be carefully studied. Based on each application, physics-based, empirical, or hybrid models might be developed in order to capture the degradation estimates [2].

Firstly, a unique definition of the degradation estimates by the help of health condition indicators is needed. In addition to that, the characteristics which highly contribute to asset degradation should be clearly defined. Based on the collected or simulated time-series data, the characteristic values should be translated into the degradation estimates [3]. Degradation evolution of the characteristics might be obtainable based on accurate knowledge regarding the physics of the failure. For many applications, understanding the physics of the failure is time- and budget-consuming. High penetration of the smart devices and sensors could highly eliminate the needs for developing the physics-based models. Indeed, as the data are becoming more available, the effectiveness of the empirical data-driven models would significantly enhance [4]. Degradation models can be viewed in two general categories as follows:

1. Offline degradation models are mainly developed based on the obtained degradation estimates over the monitoring time. Indeed, the offline degradation models are using the historical degradation estimates as the training data set to develop the models. The simplest example can be observed as finding the model which fits into the data with minimum error. The fitted model can be used for future

© Springer Nature Switzerland AG 2020   53
F. Balali et al., *Data Intensive Industrial Asset Management*,
https://doi.org/10.1007/978-3-030-35930-0_4

Table 4.1 The most important references in terms of the application of the degradation

Authors	Year	Model
Gertsbackh and Kordonskiy	1969	Application of the degradation measures in terms of sample paths to assess product reliability from the engineering point of view. They presented the Bernstein distribution, which describes the time-to-failure distribution for a simple linear model with random intercept and slope
Nelson	1981	Destructive degradation model, which have only one measurement available for each sample
Tomsky	1982	Multivariate regression models to evaluate component degradation
Amster and Hooper	1983	Simple degradation model for single, multiple, and step-stress life tests. They showed how to apply their model to estimate the central tendency of the time-to-failure distribution
Bogdanoff and Kozin	1985	The probabilistic approach to model degradation (crack length) for metal fatigue
Tortorella	1988	Markov process model for degradation data by proposing a method for estimating the parameters and testing goodness of fit
Seber and Wild	1989	Nonlinear models with dependent errors
Carey and Koeing	1991	Data analysis strategy and model fitting method to extract reliability information from degradation observation
Lu and Meeker	1993	General path models to estimate time to failure based on degradation data
Meeker and Escobar	1998	General path models with explanatory variables to estimate time to failure based on degradation data

prediction if it performs well on verification data set. Therefore, the degradation models can be developed based on the historical degradation estimates and also can be used in order to project the degradation estimates into the future [5].

2. Online degradation models mainly estimate the real-time value of the degradation based on the most recent data points. A comparison between the projected and obtained degradation estimates might reveal the needs for updating the model parameters. Online degradation models are highly beneficial for real-time monitoring purposes [6].

Degradation models let the decision-makers have an insight regarding how well the assets are working. Traditionally, lifetime test is considered as the main source of information to develop degradation models. In the era of the Internet and broad applications of the IoT devices, the needs to observe the actual failure events have been eliminated. Although lifetime information may be very helpful, obtaining lifetime data has become one of the challenges. On the other hand, technological developments lead to facing highly reliable assets which might fail zero or a few times during their useful life [7]. Therefore, the new generation of the degradation models relies on IoT devices to reflect the real condition of the asset. The output of the IoT devices can be used as the training data set for the machine learning algorithm. It is expected to reach a more accurate predictive model as the accuracy of the inputs increases [8].

Table 4.1 presents the most important references in terms of the application of the degradation models, especially for reliability assessment [9–14]. The readers may reach to each of these references to find out the detail of the proposed models and methodologies.

4.2 Applications of the Degradation Models

Degradation models may play an important role in high-level decisions regarding system maintenance and replacement policies. Degradation models can highly affect the decisions which are depending on the reliability and risk estimates. Products are usually designed with higher reliability which leads to fewer failures under the normal working conditions during the test or monitoring periods. In most of these cases, the analysts are able to extract useful information regarding the reliability of the assets based on studying the asset characteristics over time [15]. The traditional lifetime-based analyses are not an optimum decision for the recently developed products or systems due to technological enhancement. It can be highly possible that the analysts do not observe any failure during the monitoring time [16]. Indeed, due to the high penetration of the smart devices and sensors, the reliability of the units can be estimated based on the obtained accumulated test time information and assumptions regarding the distribution functions of the characteristics such as failure time. It should be noted that there is some level of uncertainties involved in these types of analyses.

Failure mechanisms can be directly linked to one or a few degradation estimates. The core of the degradation-based analyses is based on the measurement of the asset characteristic or performance data, either direct or indirect, and its relation to the presumed failure conditions. For instance, engineers can decide to consider a battery as a failed unit when its efficiency degrades up to a predefined threshold. Although the threshold value cannot be easily reached in some applications, it can be considered that engineering insights are sufficient in order to determine either a fixed deterministic threshold or the properties of the variable threshold. Degradation models can be developed based on two general categories as follows:

- In this category, the analyst has observed the overall evolution of the degradation characteristics over time. In most of the cases, this information is available for more than one sample. For instance, for the LED bulbs, the degradation estimates can be obtained by measuring the illumination of several LED lamps. Therefore, sufficient information is available to the analysts to develop a general model regarding the degradation model of the LED lamps [3].

- For some applications, observing the overall behavior of the degradation characteristics might not be practically possible. On the other hand, there might be some
  limitation regarding the number of samples under the study. In this category,
  degradation models can be developed based on extrapolating the early measurements or readings of the degradation characteristic into the future.

Degradation analysis is an efficient approach in order to assess the reliability and
risk of a product or system. Reliability assessment, risk analysis, remaining useful
life (RUL) prediction, maintenance dispatching, prognostics and health management
(PHM), and real-time monitoring are a few examples of the main applications of the
degradation models.

4.3 General Path Degradation Models

As mentioned before, the degradation estimates can be obtained based on various
approaches. In this section, it has been assumed that the time-series degradation
estimates are available for several samples. The available degradation estimates can
be fitted based on statistical approaches. Most of these approaches assume that the
failure, due to the specific degradation profile, would not occur until the predicted
degradation value crosses the prespecified critical level D or until time t_s, the
maximum safe lifetime of an asset, whichever comes first. This model has been
widely used in the degradation-based reliability assessment. The two-step general
path model is initially introduced by Lu and Meeker (1993) and Meeker and Escobar
(1998) [16, 17] and then developed by a vast number of researchers. The following
summarizes the overall steps:

1. Fit a general path model. Least squares estimation can be used to estimate the
   parameters for each path.
2. Determine the statistical distribution of each of the random parameters from the
   general path model.
3. Use the resulting distributions to solve for the time-to-failure distribution $F_T(t)$ if
   a closed-form expression exists.
4. If no closed-form expression for $F_T(t)$ exists, use the parameter distributions from
   (2) to simulate a large number N of random degradation paths.
5. To estimate $F_T(t)$, compute the proportion of random paths generated in (3) that
   cross a predetermined critical level before time t. That proportion is the estimate
   of $F_T(t)$.

$$\begin{cases} X_{i,j} = \eta_{i,j} + \varepsilon_{i,j} = \eta(t_j, \ \phi, \theta_i) + \varepsilon_{i,j} & i = 1, 2, \ldots \ldots, n \ ; \ j = 1, 2, \ldots, m \\ T_i = \min\{\mathrm{MLT}_i \ , \ \min\{t_j : \eta(t_j, \ \phi, \theta_i) \geq \eta_F\}\} \end{cases}$$

$$\varepsilon_{i,j} \sim N(0, \sigma^2{}_\varepsilon)$$

(θ_i) and $(\varepsilon_{i,j})$ are independent of each other.

$$\begin{cases} X_{i,j} & : \text{Degradation value (observed based on health indicator) for unit } i \text{ at time } j \\ \eta_{i,j} & : \text{Degradation path model for unit } i \text{ at time } j \\ \varepsilon_{i,j} & : \text{Error term for unit } i \text{ at time } j \\ t_j & : \text{ Experiment time interval} \\ \phi & : \text{ Vector of fixed-effect parameters (same for all units)} \\ \theta_i & : \text{ Vector of covariate-effect parameters for unit } i \\ \sigma^2{}_\varepsilon & : \text{Variance of error} \\ T_i & : \text{Failure time for unit } i \\ \mathrm{MLT}_i & : \text{Maximum lifetime of unit } i \\ \eta_F & : \text{Defined failure level} \end{cases}$$

Various models can be selected for the η which is supposed to explain the degradation behavior over time. The following are a few most common examples of the general path degradation process models (Fig. 4.1).

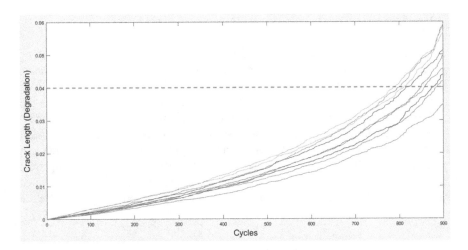

Fig. 4.1 An example of degradation profiles in terms of crack length

Examples of general degradation process models :

$$\begin{cases} \text{Linear} & : \ \eta = \phi + \theta t \\ \text{Exponential} : \ \eta = \phi . e^{\theta t} \\ \text{Power} & : \ \eta = \phi . t^{\theta} \\ \text{Logarithmic} : \ \eta = \phi + \theta . \mathrm{Ln}(t) \end{cases}$$

Distribution function of T, failure time, can be written as follows :

$$p(T_i \le t) = F_t(t) = F_T\left(t_j, \ \phi, \theta_i, \eta_F, \eta \right)$$

Suppose that the actual degradation process can be modeled as follows :

$$\eta = \phi + \theta . t \quad ;$$

$$\begin{cases} \eta(0) = \phi \\ \theta \ : \text{Degradation rate} \end{cases}$$

$$D = \phi + \theta . T \ \Rightarrow \ T = \frac{D - \phi}{\theta} \ \Rightarrow \ T = \tau(\phi, \theta, D, \eta)$$

$$\begin{cases} \text{if } \theta \sim \text{Weibull } (\alpha, \beta) & \Rightarrow F_T(T) = p\,(T < t) = \exp\left[-\left(\frac{D-\phi}{\alpha t}\right)^{\beta}\right] \\[2mm] \text{if } \theta \sim N\,(\mu, \sigma^2) & \Rightarrow \ F_T(T) = p\,(T < t) = \Phi\left(\dfrac{t - ((D - \phi)/\mu)}{\sigma t/\mu}\right) \\[2mm] \text{if } \theta \sim \text{LN}\,(\mu, \sigma^2) & \Rightarrow \ F_T(T) = p\,(T < t) = \Phi\left(\dfrac{\log\,(t) - [\log\,(D - \phi) - \mu]}{\sigma}\right) \end{cases}$$

In order to develop a more general model, a new model had been proposed by assuming that the random effect parameters, θ, or some appropriate reparameterization, $\Theta = f(\theta)$, are following a multivariate normal distribution with mean vector μ_θ and covariance-variance matrix Σ_θ. This assumption allows the analyst to summarize the information in the sample paths, without loss of substantial information, with only a mean vector and variance-covariance matrix. Estimation methods such as maximum likelihood can be applied to estimate the random parameters, which are computationally intensive. Reparameterization function, $\Theta = f(\theta)$, highly depends on the physical knowledge of the process or assumed range on some components of θ. Empirical data-driven methods can investigate the

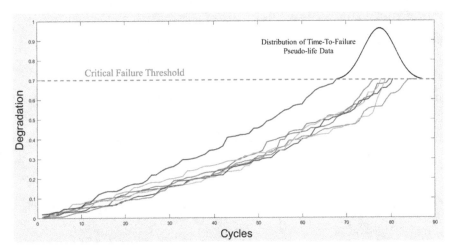

Fig. 4.2 Distribution of time-to-failure pseudo-life data based on non-destructive degradation models

process if the reparameterization is unknown. In the next step, the random effect parameters can be estimated based on two-stage method first introduced by Lu and Meeker (1993). In the first step, for each sampled unit, fit the degradation models to the sample path and obtain Stage 1 estimates. In Step 2, obtained estimates of the degradation model parameters should be combined to produce ϕ ,μ_θ, and Σ_θ.

4.4   Non-destructive Degradation Models

In this case, multiple degradation measurements can be obtained for each sample at each point of the time. Given the failure threshold, degradation measurements can be extrapolated to the point where the failure will occur. Extrapolated failure times can be viewed as traditional lifetime data in order to be analyzed in the same way. Extrapolated failure times are usually called pseudo-life data. This indicates that the analysts did not wait to observe the physical failures of the system. These data points have been collected as the results of either observing the whole degradation profile or extrapolating the degradation estimates into the future with some levels of uncertainty. Figures 4.2 and 4.3 depict the distribution of time-to-failure pseudo-life data based on non-destructive degradation models when the degradation profiles have been obtained or extrapolated for several samples over time, respectively.

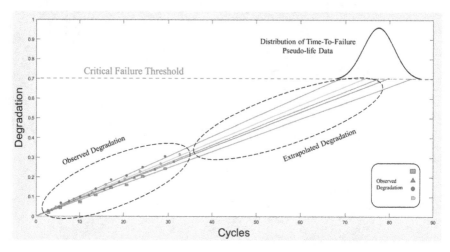

Fig. 4.3 Distribution of time-to-failure pseudo-life data based on extrapolated non-destructive degradation models

4.5 Non-destructive Degradation Predictive Models

When the overall degradation estimates have been observed, the next step would be estimating the parameters of the selected model to be considered as the degradation model. If the degradation estimates only recorded or observed for a few periods, which is not very close to the threshold value, the analyst should extrapolate the degradation estimates by assuming a degradation model. Linear, exponential, power, or logarithmic models are a few examples of the most commonly used models for the degradation models, while y represents the performance characteristic, x represents the time, and a and b are model parameters which must be estimated given a data set [16].

$$y = ax + b \qquad\qquad\qquad \text{Linear}$$

$$y = b.e^{ax} \qquad\qquad\qquad \text{Exponential}$$

$$y = b.x^{a} \qquad\qquad\qquad \text{Power}$$

$$y = a. \ln{(x)} + b \qquad\qquad \text{Logarithmic}$$

$$y = a - \frac{b}{x} \qquad\qquad\qquad \text{Lloyd} - \text{Lipow}$$

The following is an example of the point estimation of the parameters for the viscosity ratio (10^{-3} m^2/s) of the transformer oil over time for the five samples. In this example, we assumed that the velocity ratio can be defined as nominal velocity over velocity at each point of the time. It should be considered that each analyst might have different health indicator index. In the next step, it has been assumed that

Table 4.2 Input characteristic data of oil properties for each test unit

Cycle	Unit 1	Unit 2	Unit 3	Unit 4	Unit 5
100	15	12	16	15	14
200	20	16	21	18	19
300	22	21	25	23	21
400	28	26	28	27	26
500	30	29	31	31	30

Table 4.3 Estimated exponential parameters for each test unit

Parameters	Unit 1	Unit 2	Unit 3	Unit 4	Unit 5
a	0.0016	0.0020	0.0015	0.0018	0.0017
b	13.7710	10.7510	15.1570	12.9250	12.5600

Table 4.4 Cycles-to-failure for each test unit based on the exponential model

	Unit 1	Unit 2	Unit 3	Unit 4	Unit 5
Cycles-to-failure	481	513	488	495	512

the health index can be explained by the exponential function. Therefore, the purpose would be estimating the parameters, a and b, of the exponential function for each test unit by using the nonlinear regression analysis methods. Based on the engineering insight, it has been concluded that failure occurs when the ratio reaches to 30. As a result, failure time for each test unit can be obtained by solving the equation ( ). This time to failure can be analyzed using traditional life data analysis to obtain metrics such as the probability of failure and reliability (Tables 4.2, 4.3, and 4.4).

$$y = b.e^{ax} \qquad\qquad \text{Exponential}$$

$$\overset{\text{yields}}{\longrightarrow} x = \frac{\ln(y) - \ln(b)}{a}$$

The uncertainty in the prediction is highly dependent on the number of data points which the information is available. It is expected to reach more robust models as the size of the input data increases. The analyst might try more than one distribution in order to find the best fit to the available data before extrapolating into the feature. Mean squared error (MSE), root mean squared error (RMSE), likelihood R-squared, R-squared adjusted, and coefficient of variation (CV) are a few examples of the statistics which are commonly used to compare the performance of various models. In this example, the model performances are compared based on RMSE value. It might be possible that more than one statistics is needed in order to draw a valid conclusion. As Table 4.5 presents, for all the test units except unit 3, the linear model has the lowest RMSE value which indicates a lower error in terms of the predicted versus actual values. Therefore, the analyst probably ends up with using the linear model for all the units except unit 3. Power model would have the lowest RMSE for unit 3. It should be noted that five data points, which are available for each sample,

Table 4.5 RMSE values for each test unit based on various models

	Unit 1	Unit 2	Unit 3	Unit 4	Unit 5
Linear	1.100	0.632	0.796	0.483	0.796
Exponential	1.440	1.420	1.370	0.804	0.961
Power	1.200	0.803	0.158	1.180	1.200
Logarithmic	1.670	1.790	0.790	2.010	1.840

Table 4.6 Cycles-to-failure based on the RMSE criteria

	Unit 1	Unit 2	Unit 3	Unit 4	Unit 5
Model	Linear	Linear	Power	Linear	Linear
a	0.038	0.044	0.413	0.041	0.039
b	11.6	7.6	2.36	10.500	10.300
Cycles-to-failure	484	509	472	476	505

might not be sufficient to generalize the results for the similar units. As mentioned before, the uncertainties could be improved as the number of data points increases (Table 4.6).

4.6   Destructive Degradation Models

Destructive degradation analysis refers to the type of analysis which the product would destroy during obtaining the degradation information. Consequently, the degradation information may be obtained by measuring the characteristics of several samples at different points of the time. Indeed, destructive degradation models are very close to the accelerated life test analysis (ALTA). The main difference between the destructive and non-destructive degradation models is in the random variable definition. Non-destructive degradation models consider the time to failure as a random variable, while the destructive degradation models study the degradation measurements as a random variable. Figure 4.4 presents the degradation, in terms of voltage deterioration, as a random variable for destructive tests.

The destructive analysis includes the statistical distribution analysis regarding the performance of the degradation or characteristics over time. Weibull, normal, log-normal, and exponential are a few examples of the most commonly used distributions for asset performance characteristics. Most of the studies considered that the location or log of location parameter will change over time, while shape parameter remains the same.

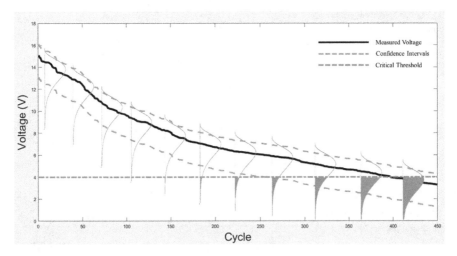

Fig. 4.4 Degradation as a random variable for destructive tests

4.7   Destructive Degradation Predictive Models

As mentioned before, the core of the analysis assumes that the location or log of location will change over time. Based on the selected degradation model, the parameter of the model can be estimated by applying the maximum likelihood estimation (MLE) principles. All the abovementioned models, such as linear, exponential, power, and logarithmic models, can all be used for the destructive degradation prediction models as well.

Assume that a degradation profile is following a normal distribution to explain the degradation behavior at each point of time. Therefore, standard deviation, σ, would remain a constant value, while mean, μ, would change over time. Therefore, the cumulative density function (CDF) of the degradation, $\eta(t)$, can be estimated as follows, while η_F presents the critical threshold value which the failure occurs.

$$p(\eta(t) \leq \eta_F) = \phi\left(\frac{\eta_F - \mu(t)}{\sigma}\right)$$

If we assume that the degradation model can be explained by the aim of linear models, $\mu(t)$ can be estimated based on the estimates of the parameters of the linear models as follows:

$$\mu(t) = a + b.t \qquad \text{(linear model)}$$

Example A company is monitoring power consumption of an electric drive over a period of 5 months. As the manufacturer suggested, maintenance actions are needed on average every 6 months, due to the degradation in terms of excessive electrical

Table 4.7 Normalized power consumption as the measured degradations over the 5 months

Month 1	Month 2	Month 3	Month 4	Month 5
1.20	1.35	1.33	1.45	1.48
1.22	1.29	1.42	1.46	1.51
1.25	1.43	1.41	1.49	1.57
1.25	1.58	1.54	1.54	1.61
1.28	1.52	1.58	1.52	1.69
1.35	1.49	1.63	1.59	1.74
1.45	1.59	1.63	1.67	1.75
1.46		1.64	1.71	1.75
1.47			1.78	1.78
1.51				

Table 4.8 The value of the estimated parameters for each month

	Month 1	Month 2	Month 3	Month 4	Month 5
μ	1.34	1.46	1.52	1.58	1.65
σ	0.11	0.11	0.11	0.11	0.11

Table 4.9 Reliability estimates

T	$R(t) = p(\eta(t) \le 1.8)$
1	1
2	0.9995
3	0.9957
4	0.9750
5	0.9015
6	0.7318

power consumption. It has been assumed that the drive needs maintenance when the power ratio reaches to 1.8. Table 4.7 presents the normalized power consumption as the measured degradations over the 5 months. The goal is to estimate the reliability and probability of failure of the same drive given this information (Tables 4.8 and 4.9).

Now, we want to estimate the parameters using the linear degradation model and the normal distribution function. Recall the linear degradation model as follows (Fig. 4.5):

$$\mu(t) = a + b.t \qquad (\text{linear model})$$
$$\mu(t) = 1.288 + 0.074.t$$
$$R(t) = p(\eta(t) \le 1.8) = \phi\left(\frac{1.8 - \mu(t)}{0.014}\right)$$

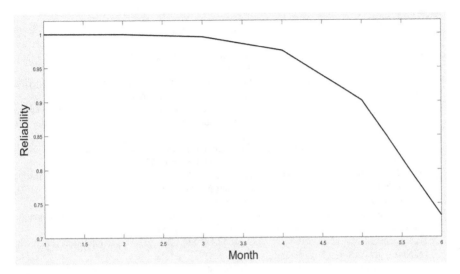

Fig. 4.5 Estimated reliability based on the destructive degradation models

4.8   Concluding Remarks

Principles of the degradation mechanism and general categories of the predictive degradation models have been explained in detail in this chapter. Degradation estimates must be obtained before developing predictive degradation models since the degradation estimates should be fed into the predictive algorithms for training purposes. Degradation estimates can be measured, collected, simulated, or estimated based on various principles, which highly depend on the application. In this chapter, it has been assumed that the degradation estimates are available to the analysts. The next step would be developing the prediction degradation model in order to have an insight regarding the future condition of the assets. Degradation models can be categorized based on either destructive or non-destructive test to collect information regarding the performance indicator or asset characteristics. The main target of the non-destructive degradation-based model is estimating the distribution of the time-to-failure data estimated based on the first time which degradation profile crosses the threshold. Therefore, the focus is more toward the time to event approaches. For the destructive degradation model, the main purpose is to find the distribution of the degradation at each point of the time. The main reason is that there are a limited number of measurements for each sample unit. In the next step, reliability can be estimated either based on the time-to-failure or degradation distribution. For each of these approaches, a numerical example has been presented to further clarify the detail.

References

1. A. Van Horenbeek, L. Pintelon, A dynamic predictive maintenance policy for complex multi-component systems. Reliab. Eng. Syst. Saf. **120**, 39–50 (2013)
2. W. Liao, Y. Wang, E. Pan, Single-machine-based predictive maintenance model considering intelligent machinery prognostics. Int. J. Adv. Manuf. Technol. **63**(1-4), 51–63 (2012)
3. D. Wang, Q. Miao, M. Pecht, Prognostics of lithium-ion batteries based on relevance vectors and a conditional three-parameter capacity degradation model. J. Power Sources **239**, 253–264 (2013)
4. W. Ahmad, S.A. Khan, J. Kim, A hybrid prognostics technique for rolling element bearings using adaptive predictive models. IEEE Trans. Ind. Electron. **65**(2), 1577–1584 (2018)
5. D.A. Tobon-Mejia, K. Medjaher, N. Zerhouni, G. Tripot, A data-driven failure prognostics method based on mixture of Gaussians hidden Markov models. IEEE Trans. Reliab. **61**(2), 491–503 (2012)
6. K. Efthymiou, N. Papakostas, D. Mourtzis, G. Chryssolouris, On a predictive maintenance platform for production systems. Procedia CIRP **3**, 221–226 (2012)
7. N. Gorjian, L. Ma, M. Mittinty, P. Yarlagadda, Y. Sun, A review on degradation models in reliability analysis, pp. 369–384
8. Z. Ye, Y. Wang, K. Tsui, M. Pecht, Degradation data analysis using Wiener processes with measurement errors. IEEE Trans. Reliab. **62**(4), 772–780 (2013)
9. K. Smith, E. Wood, S. Santhanagopalan, G. Kim, Y. Shi and A. Pesaran, No title, Predictive models of li-ion battery lifetime
10. X. Zhang, E. Gockenbach, Asset-management of transformers based on condition monitoring and standard diagnosis. IEEE Electr. Insul. Mag. **24**(4), 26–40 (2008)
11. J. Jalbert, R. Gilbert, Y. Denos, P. Gervais, Methanol: A novel approach to power transformer asset management. IEEE Trans. Power Del. **27**(2), 514–520 (2012)
12. J. Tang, T. Su, Estimating failure time distribution and its parameters based on intermediate data from a Wiener degradation model. Nav. Res. Logist. **55**(3), 265–276 (2008)
13. M. Giorgio, M. Guida, G. Pulcini, An age-and state-dependent Markov model for degradation processes. IIE Trans. **43**(9), 621–632 (2011)
14. L.A. Escobar, W.Q. Meeker, D.L. Kugler, L.L. Kramer, Accelerated destructive degradation tests: Data, models, and analysis. Math. Stat. Methods Reliab, 319–337 (2003)
15. S. Alaswad, Y. Xiang, A review on condition-based maintenance optimization models for stochastically deteriorating system. Reliab. Eng. Syst. Saf. **157**, 54–63 (2017)
16. W.Q. Meeker, L.A. Escobar, *Statistical Methods for Reliability Data* (Wiley, New York, 2014)
17. W.Q. Meeker, L.A. Escobar, A review of recent research and current issues in accelerated testing. Int. Stat. Rev./Revue Internationale de Statistique **61**, 147–168 (1993)

Chapter 5
IoT Platform: Smart Devices, Gateways, and Communication Networks

5.1 Smart Devices

Dozens of smart devices can easily be found by a glance view of the objects surrounding us. Smartphones, laptops, tablets, and smart watches are the first couples of notable smart device examples due to their broad applications in daily human life [1]. In favor of IoT, a smart device is a network-enabled electric device (Thing) connected to other devices, networks, or clouds via various wire and wireless protocols and standards such as Bluetooth, NFC, 3G, 4G, etc. which is able to operate to some extent autonomously and interactively [2, 3]. Smart devices are mostly able to exchange data with other devices and other operating systems such as manufacturers, operators, service companies, etc. Smart devices usually are a set of electronic hardware and software components which are capable of connecting, sharing, and interacting with other devices and networks and, to some extent, have the computation power to perform early analysis at the component level [4]. Smart devices are designed to be used in three main environments as the physical world, human-centered environment, and distributed computing environment. The exchanged data from the component toward the upper system level can precisely be analyzed to optimize the decision-making process [5].

There are several currently existing devices which are not smart enough to be able to interact with other smart devices through an IoT platform. Although the current trend is toward being smarter, the trade-off between the cost and level of being smart should always be considered carefully. A business cannot easily conclude that replacing a currently existing asset with a new smart one can have the actual benefits to the business [6]. There are vast major questions which must be answered before making these high-level decisions. A few of these questions are presented as follows:

- What type of information (data) is needed which currently do not exist?
- What are the applications of the data?
- What types of benefits can be added up to the business by accessing those data?
- What is the ultimate outcome given that information?

© Springer Nature Switzerland AG 2020
F. Balali et al., *Data Intensive Industrial Asset Management*,
https://doi.org/10.1007/978-3-030-35930-0_5

- What are the effects of new assets on operating costs such as bandwidth costs?
- What is the expected improvement in the availability and reliability of the system?

There are a variety of IoT smart devices in several sizes and qualities capable of performing numerous functions. Most of the time, these devices have low-powered and low-cost functionality to be able to perform the basic analysis at the component level. The storage capacity and computing power are limited for smart components. The burden of the analysis is mostly on edge and cloud levels [7].

Majority of the IoT smart devices are employing the low-cost microprocessors, microcontroller, or central processing unit. The storage can be an external device or an onboard microchip. Most of the time, microprocessors are coupled with an external storage unit to enhance their flexibility. On the other hand, an onboard storage chip is usually attached to the microcontrollers to optimize the cost and energy consumption as well as faster response time. Smart devices can be powered in three popular ways as alternate current (AC), direct current (DC), and battery. AC can provide power as much as needed by the units. It should be noted that an AC/DC converter might be needed to be embedded to the device due to the presence of line voltage. In case of the available power source, DC might be a good option because of the required low-cost and simple DC/DC converter. For many applications, the battery is the best option if the size and location of the battery are optimized and chosen based on the requirements. A variety of attributes regarding the choice of the batteries should be well-thought-out [8]. A few examples are nominal output voltage, energy capacity, temperature, operating mode, self-discharge rate, power drop, etc. There are various types of smart sensors for each application. The following is a summary of the most commonly used sensors [9]:

- Acoustic, sound, vibration
    - Geophone
    - Sound locator

- Automotive
    - Airbag sensors
    - Speed sensor

- Chemical
    - Breathalyzer
    - Carbon dioxide sensor
    - Smoke detector

- Electric current, electric potential, magnetic, radio
    - Current sensor
    - Voltage detector

- Environment, weather, moisture, humidity

- – Air pollution sensor
- Flow, fluid velocity
  - – Mass flow sensor
- Ionizing radiation, subatomic particles
  - – Cloud chamber
- Navigation systems
  - – Air speed indicator
- Position, angle, displacement, distance, speed, acceleration
  - – Tilt Sensor
- Optics, light, imaging, photon
  - – Infrared sensor
- Pressure
  - – Pressure gauge
- Force, density, level
  - – Torque sensor
- Thermal, heat, temperature
  - – Thermometer
- Proximity, presence
  - – Thermometer

Figure 5.1 depicts an example regarding the smartwatches such as Apple or Fitbit watches which track the physical activities of the person. A part of the measurements

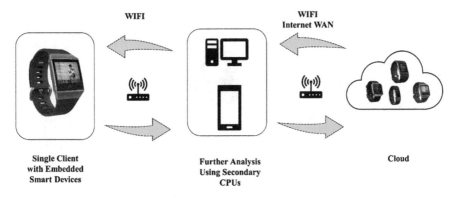

Fig. 5.1 An example of the development of smart devices

such as heart bit is instantly available at the component level. The other layer such as a cellphone or laptop extracts the information sent by smartwatch through the Wi-Fi connection. At this level, a few analyses such as physical activities or sleeping trends can be available to the users. It should be noted that some features such as the theme and background of the watch can be managed through the phone or laptop. Therefore, the type of connection is interactive. At the highest level, information and analytics can be transferred to the central cloud for further analysis. For instance, the analytics at this level can compare the persons' activity level with other similar groups with the same health conditions [10].

5.2   Device Security

Security is one of the most concerning issues of the smart devices. Although the interconnection is the key for an IoT platform, any connection outside of the defined boundaries of the system might be a danger [11]. It should be noted that security is needed at every point of the platform. It starts with the component level such as sensors toward the high-level analysis at edge and cloud. There are several policies and standards regarding a secure connection between the complements of an IoT platform. IoT smart devices might bring convenience and business advantages, but security is the first concern which has to be considered before implementing the smart devices into the network [12]. A look at the Open Web Application Security Project's (OWASP), "Internet of Things Top 10," reveals such a weakness, concern, or lack of security and weak authentication mechanisms. Figure 5.2 presents a schematic view of the security of an IoT-based network in either component or system levels.

Fig. 5.2 Schematic view of the security of an IoT-based network in either component or system levels

As a real-world example of the importance of security of a network, Counter Threat Unit Research Team on the Security of Smart Devices stated "a real-world example of these security risks came courtesy of TRENDnet. In February 2014, the US Federal Trade Commission (FTC) approved a settlement with TRENDnet over charges that while the company claimed its SecurView cameras were secure, they had faulty software that left them open to online viewing, and in some cases listening, by anyone with the cameras' Internet address. As part of the settlement, the FTC ordered TRENDnet to establish a comprehensive information security program to address risks that could result in unauthorized access or use of the company's devices and mandated TRENDnet to obtain third-party assessments of its security programs every 2 years for the next two decades."

By putting the smart devices into the network of an organization, new ways of connection between the smart devices and web-based services or cloud providers are established. In some cases, there might be a few devices with multiple homes, which means belonging to more than one network. In these cases, security and bypassing the firewalls are concerning. Indeed, the most important issue would be the level of trust [13]. All of the registered devices for a network should be trustable enough to be able to communicate and exchange the information. Therefore, due to the interconnection between the devices and cloud providers, the issue is not anymore only related to the single units. It might be possible that a third party can have access to a considerable portion of the information and analytics given access to a single asset. A secure IoT platform must contain the following [14–16]:

- Secure startup and boot is necessary for all the smart devices to prevent any unauthorized booting of the IoT devices by boot access preventive policies. This step is in charge of examining any of the hardware and software to make sure none of them are modified without passing the approval procedure for the modifications.
- The software update should perform securely with respect to the newly released version of terms and conditions. This might enable a unit to have access to the credential information due to the acknowledged and accepted terms and conditions.
- Data access procedure must be clearly defined in advance to follow the security levels. One of the currently existing challenges regarding smart IoT system is defining the entities who should have access to various levels of information.
- Communication network should be secured in a way that the authorization and authentication of the users be confirmed for each activity.
- Firewalls and preventive cyberattack procedures must be implemented for the entire network.
- Policy management should be frequently reviewed and updated to guarantee a safe communication network.

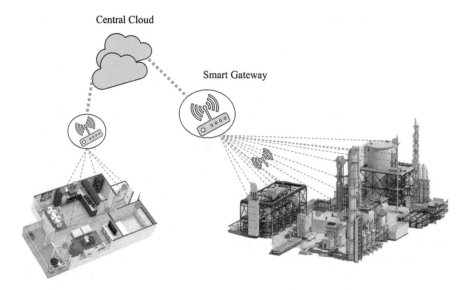

Fig. 5.3 Schematic view of sensors, smart gateway, and cloud

5.3   Smart Gateway

Every IoT platform needs to create a connection between the smart devices and cloud providers so that the data can be sent back and forth between them. One essential for every IoT platform is the gateways. Indeed, gateways act as a bridge between the sensors and the cloud. Sensors are directly communicating with the gateways. Indeed, gateways translate and transfer the information to the central cloud [17]. Your question might be: Why should sensors data not directly send the data to the cloud? As an example, consider a sensor in a remote area which needs a long-range connection to be able to talk to the cloud [18]. From a business point of view, the long-range connection means more power consumption and more cost. Another reason could be the required bandwidth which makes all the smart devices able to communicate with the cloud. Smart gateways can help smart devices to only transmit the data in a short range. As a result, the lifetime of the sensors will be improved. Smart gateways are able to backhaul the data to the cloud through a single high bandwidth connection [19]. Figure 5.3 presents a schematic view of sensors, smart gateway, and cloud.

All the sensors of an IoT platform might not be the same in terms of size, functions, and manufacturers. Therefore, each of them might use various communication protocols. A few examples of the protocols include LPWAN, Wi-Fi, Bluetooth, and Zigbee. Gateways can communicate with sensors with various protocols and then translate all the collected information into a single protocol such as MQTT to be able to transmit to the cloud [20]. This also increases the security of the system since the sensors and smart devices are only connected to the smart gateways.

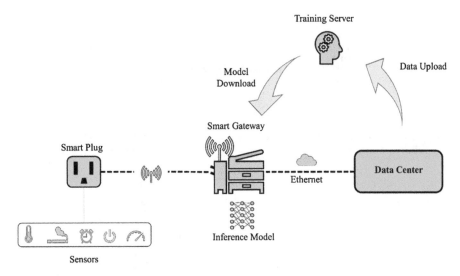

Fig. 5.4 The overview of an IoT framework system

The Internet connection exists between the smart gateways and central cloud. It should be noted that this requires high-level security requirements for smart gateways since all the information collected and analyzed based on the sensor measurements are accessible at the gateway. Figure 5.4 depicts an overview of an IoT framework system [21].

Most of the time, it is not necessary to transmit the whole portion of the data. For instance, nobody is interested in the ambient temperature profile when the production system is completely shut down. Therefore, data filtration and preprocesses can be considered as other advantages of smart gateways [22]. In addition to that, time is always important in an IoT platform especially for the critical components which need immediate attention. For instance, sometimes the transmission time from the sensor to the cloud is much longer than the period which the sensors and actuators should take action. Consequently, smart gateways are able to reduce the latency which is an important feature especially for the time-critical application. Smart gateways enable the systems to take actions at the smart gateway level instead of the central cloud [23].

5.4   Communication Types

From the device point of view, various types of communication exist.

- *Device to Device*
  Direct communication between the devices without traversing any station. Devices should be able to communicate with each other autonomously without

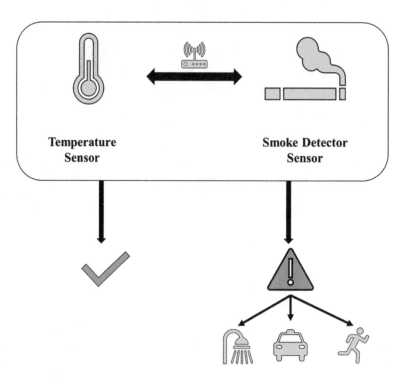

Fig. 5.5 An example of device-to-device communication

any centralized controller to gather, share, and transfer the information. The IoT smart devices require intelligent protocols in order to achieve smart and safe device-to-device communication. Device-to-device communication is an essential part of the IoT platform to main the quality of the service above a certain level. In this type of communication, the centralized controller is not needed anymore which leads to more spectrum- and energy-efficient network with optimized traffic load. Device-to-device communication is applicable to real-time controlling and monitoring. Raspberry Pi and Intel Edison's devices are a couple of examples which are able to perform the device-to-device communications. Figure 5.5 presents an example of device-to-device communication for a smart house. If any of the temperature or smoke detector sensors identify a dangerous situation, they are able to communicate with each other to find critical status of the other sensors. For instance, high temperature and a high level of carbon monoxide and dioxide are probably a sign of potential fire. Evacuation alarm can be sounded to notify the tenants after the device-to-device interaction of the smart sensors. In addition to that, smart sensors and actuators might activate the fire sprinkler system and automatically notify the emergency departments.

- *Device to Gateways*

  As it has been discussed above, devices are connected to the smart gateways and exchange the information in terms of data. Smart gateways are able to unify the outputs of various types of sensors and make it ready to push toward the central cloud. Sensor measurements are directly sent to the smart gateways before performing any type of filtration or analysis. Each application might have one or several smart gateways.

- *Device to Cloud*

  Rarely a device is directly connected to the cloud due to the cost, security, and latency. If a device is measuring a very critical variable which all the measurements should be recorded, it might be possible that the output directly pushes to the cloud. In most of the times, smart gateways are a connection between the devices and cloud.

5.5 Network Service

In today's market, there are various network service providers. Each service might be tailored for a specific application or it might be applicable to general businesses. There are a variety of factors which should be considered for selecting the network service providers [24, 25].

- *Cost*: Just like many other applications, cost is one of the most critical factors for selecting the right network service providers. Low-cost services usually should be able to offer lower prices than cellular, Wi-Fi, and satellite.
- *Power*: Low-powered devices mostly last for a longer period of time. It should always take into account that the network bandwidth can highly affect the power consumption of the devices.
- *Reliability*: A reliable network service provider should be able to offer an adequate and stable service. It should be noted that inadequate or unstable service might burden excessive and unexpected cost to an organization. For instance, interruption or large delays in the connection of a few sensors to the gateways might cause a very harmful situation. A system which highly relies on the early warning system might already face the undesirable scenario in case of the unreliability of the IoT network service providers.
- *Bidirectional Communication*: IoT network service providers should be able to offer bidirectional communication in order to exchange the information and data between the smart sensors, actuators, smart gateways, cloud, etc.
- *Security*: Security is one of the most important features in selecting network service providers. Security should be carefully considered at the component level (either built-in or added function), communication and data transfer, and gateway and cloud levels.
- *Long Range*: Based on the application, 15 miles for rural and 5 miles for urban areas can be considered as long-range service.

5.6   Data Network Security

As mentioned earlier in the section of this chapter, security is one of the most concerning criteria for each network which integrated the smart devices and communication networks. Network security includes a variety of policies to manage and monitor the access of the authorized and unauthorized users. In other words, users must obtain network administrator authorization in order to access a particular data. In most of the applications, administrators assign authentication information, such as username and password, to each user who wants access to information and programs. Network security is covering the public and private networks. The simplest way of protecting a network is assigning a unique name and password to the network. Wi-Fi username and password can be considered as one of the basic levels of the network security. Obviously, the needed actions regarding protecting the network security for an organization are beyond assigning the username and password for the network and smart devices. Once the network approves the authentication of a user, the firewall enforces the access policies which define what level of information can be accessible by the authorized user. Security management is different for each situation. Obviously, larger businesses need more advanced security management in order to prevent malicious attacks from hacking and spamming. The levels of security management within various sectors of an organization are also different. Security is a key element in an IoT-based network for all the levels starting from the components up to the edge and central cloud.

5.7   Concluding Remarks

The current era can be called the era of smart devices. The recent trend over the last few decades is toward being smarter. This chapter reviews the most important principles of the smart devices and the features which make a device smart. A careful analysis is needed before updating the currently existing platform based on the IoT principles. This chapter also studied the role of smart gateways, the architecture of an IoT platform, the detail of the communication types, and the most important features of network service providers for an IoT platform.

References

1. M. Weiser, The computer for the twenty-first century. Sci. Am. **265**(3), 94–104 (1991). https://doi.org/10.1038/scientificamerican0991-94
2. S. Poslad, *Ubiquitous Computing Smart Devices, Smart Environments and Smart Interaction.* PP. 642–644. (Wiley, Sirirajmedj Com 67.10, 2009). ISBN 978-0-470-03560-3

3. A. Stisen et al., Smart devices are different: Assessing and mitigating mobile sensing hetero-geneities for activity recognition. Proceedings of the 13th ACM conference on embedded networked sensor systems. ACM, 2015
4. G. Suarez-Tangil et al., Evolution, detection and analysis of malware for smart devices. IEEE Commun. Surv. Tutorials **16.2**, 961–987 (2014)
5. J. Harwood et al., Constantly connected–the effects of smart-devices on mental health. Comput. Hum. Behav. **34**, 267–272 (2014)
6. M. McTear, Z. Callejas, D. Griol, *Conversational Interfaces: Devices, Wearables, Virtual Agents, and Robots. The Conversational Interface* (Springer, Cham, 2016), pp. 283–308
7. S. Nihtianov, A. Luque (eds.), *Smart Sensors and MEMS: Intelligent Sensing Devices and Microsystems for Industrial Applications.* (Woodhead Publishing, Duxford, 2018)
8. A. Das, N. Borisov, M. Caesar, Do you hear what i hear?: Fingerprinting smart devices through embedded acoustic components. Proceedings of the 2014 ACM SIGSAC conference on computer and communications security. ACM, 2014
9. S. Bennett, A history of control engineering 1930–1955. London: Peter Peregrinus Ltd. on behalf of the Institution of Electrical Engineers. ISBN 0-86341-280-7<The source states "controls" rather than "sensors", so its applicability is assumed. Man, (1993)
10. Hong Li et al., WiFinger: talk to your smart devices with finger-grained gesture. Proceedings of the 2016 ACM international joint conference on pervasive and ubiquitous computing. ACM, (2016)
11. Min Gyu Chung et al., Method for protecting contents, method for sharing contents and device based on security level. U.S. Patent No. 8,949,926. 3 Feb 2015
12. A. Sinha, Cloud based mobile device security and policy enforcement. U.S. Patent No. 9,119,017. 25 Aug 2015
13. Hyok-Sung Choi et al., Method and apparatus for handling security level of device on network. U.S. Patent No. 8,875,242. 28 Oct 2014
14. F. Hoornaert, M. Houthooft, Field programmable smart card terminal and token device. U.S. Patent No. 8,949,608. 3 Feb 2015
15. H. Rathore et al., DTW based authentication for wireless medical device security. 2018 14th international wireless communications & mobile computing conference (IWCMC). IEEE, 2018
16. R.J. Delatorre et al., Network based device security and controls. U.S. Patent No. 9,055,090. 9 Jun. 2015
17. M. Aazam, Eui-Nam Huh, Fog computing and smart gateway based communication for cloud of things. Future Internet of Things and cloud (FiCloud), 2014 international conference on. IEEE, 2014
18. M. Aazam, Pham Phuoc Hung, Eui-Nam Huh, Smart gateway based communication for cloud of things. Intelligent sensors, sensor networks and information processing (ISSNIP), 2014 IEEE Ninth International Conference on. IEEE, 2014
19. L. Catarinucci et al., An IoT-aware architecture for smart healthcare systems. IEEE Internet Things J **2**(6), 515–526 (2015)
20. Te-Sheng Chen et al., Smart gateway, smart home system and smart controlling method thereof. U.S. Patent No. 9,547,980. 17 Jan 2017
21. C. Razafimandimby et al., A bayesian and smart gateway based communication for noisy iot scenario. Computing, networking and communications (ICNC), 2017 international conference on. IEEE, 2017
22. I. Tcarenko et al., Smart energy efficient gateway for Internet of mobile things. Consumer communications & networking conference (CCNC), 2017 14th IEEE Annual. IEEE, 2017.
23. Tein-Yaw Chung et al., Design and implementation of light-weight smart home gateway for Social Web of Things. Ubiquitous and future networks (ICUFN), 2014 sixth international conference on. IEEE, 2014
24. N. Mishra, In-network distributed analytics on data-centric IoT network for BI-service applications. Int. J. Sci. Res. Comput. Sci. Eng. Inf. Technol, ISSN, 2456–3307 (2017)
25. Qian Wang et al., CS-Man: Computation service management for IoT in-network processing. Signals and systems conference (ISSC), 2016 27th Irish. IEEE, 2016

Chapter 6
Data Features

6.1 Data Analysis

As mentioned in the previous chapters, the main characteristic of an IoT-based platform is the connectivity between the components of the system. Components can be connected to each other within a smart platform and gateway to obtain the data close to real time. It should be considered that the connectivity is the main difference between the modern and traditional architectures. Analysts should always consider as added value procedures due to all the IoT-based activities. An IoT-based platform with various smart connected products and enormous collected real-time data might fail if the added values as the gain of updated structure cannot rationalize the investment. The ability of an efficient IoT-based platform reveals after performing various analyses on the collected data. Therefore, data analytic techniques should be able to directly extract the most useful information out of the collected data to enhance the added value of the platform.

Data analysis is a series of procedures which should be performed on the raw data to extract useful information [1]. Raw data collected from the sensor measurements cannot inherently reveal any useful information. These processes include data cleansing, filtration, transformation, etc. A variety of statistical data analysis techniques can be applied to the processed data to drive a conclusion or decide regarding the system behavior [2]. In other words, data analysis tries to convert the raw data into the useful information which facilitate the decision-making processes. Raw sensor measurements are the output of the data collection which should be further analyzed to answer the questions, reveal abnormal situations or trends, test hypothesizes, and ultimately report and store the most useful information [3]. Statistician John Tukey defined data analysis in 1961 as "Procedures for analyzing data, techniques for interpreting the results of such procedures, ways of planning the gathering of data to make its analysis easier, more precise or more accurate, and all the machinery and results of (mathematical) statistics which apply to analyzing data" [4, 5]. In the statistical area, data analysis divides into two main categories as

© Springer Nature Switzerland AG 2020
F. Balali et al., *Data Intensive Industrial Asset Management*,
https://doi.org/10.1007/978-3-030-35930-0_6

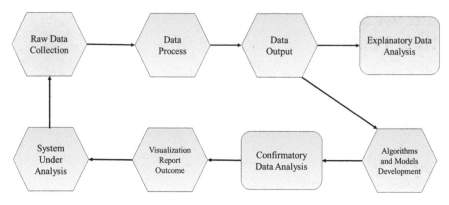

Fig. 6.1 A big picture of data analysis processes

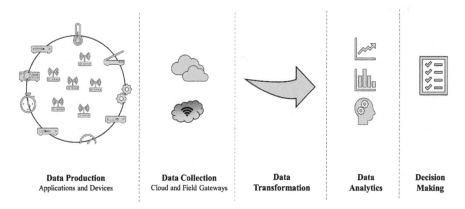

Fig. 6.2 An example of the architecture of the data stream in an IoT platform

explanatory data analysis (EDA) and confirmatory data analysis (CDA). Undiscovered feature or trend should be revealed by applying the EDA techniques. The hypothesis can be tested using the CDA principles. Figure 6.1 depicts a big picture of the data analysis processes. Figure 6.2 presents an example of the architecture of the data stream in an IoT platform.

6.1.1   Qualitative and Quantitative Analysis

Data analysis can be categorized as qualitative and quantitative analysis. These two categories underlay different fundamentals and principles while still sharing a number of the same objectives [6]. These techniques can be applied from the early stages of data collection toward the final steps of data analysis. They can be used separately or concurrently since each of them are attached to the bias and error. The

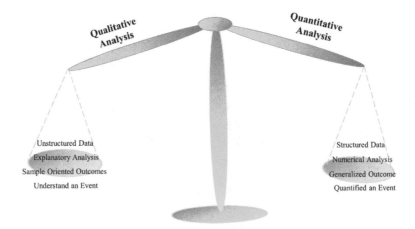

Fig. 6.3 Quantitative and qualitative analysis

main difference between these two groups is in data collection, data analytics, and outcomes [7]. Figure 6.3 depicts the main differences between qualitative and quantitative analysis.

Quantitative Analysis The main objective of the quantitative analysis is quantifying the data. In other words, quantitative analysis refers to numerical statistical analysis. In this approach, the frequencies should be assigned to each of the features. The nature of the features should be measurable. Quantitative data can be classified as discrete or continuous random variables. Analyses are based on the statistical methods to reach an outcome. As long as a valid sampling and modeling approach is being applied, outcomes can be generalized to a larger population [8]. Quantitative techniques are more object-oriented which means try to understand the occurrence of an event and then describe them using the statistical models. Quantitative methods might not be able to find out the reason for the occurrence of an event. Therefore, the two categories can be applied at the same time to clearly understand the context of a system and then numerically study the system behaviors [9, 10].

Qualitative Analysis The main objective of the qualitative analyses is reaching an insight and understanding of the reasons for a system. The results of the qualitative analysis can guide the analyst to develop a more applicable and reasonable hypothesis for qualitative analysis [11]. The nature of the qualitative analysis is more toward explanatory and investigatory analysis. This indicates the main disadvantage of qualitative analysis. It means that the outcomes of the qualitative analysis might not be directly generalized to the whole population with a certain level of uncertainty. The reason is that no hypotheses usually being tested to study that the outcomes are statistically significant or based on the random chance [12]. In general, the results of the qualitative analysis are more about the in-depth understanding of a context. In this type of analysis, no attempt is needed to assign frequencies to any of the features [9].

Fig. 6.4 Parametric and nonparametric tests

6.1.2   Parametric and Nonparametric Analysis

Assume a sample is drawn from a population with unknown parameters. However, some estimates can be made based on the sample about a few parameters such as mean and variance [13]. The calculated parameters are called "statistic" which are an estimate of the unknown parameter of the population given the drawn sample. The parametric statistical analysis relies on assumptions regarding the shape or parameters of the distribution. Nonparametric methods depend on none or a few assumptions regarding the parameters of the population which the sample is drawn [14]. Nonparametric techniques are not based on the parametrized families of distribution functions. Indeed, they are distributed free or belong to a distribution with unspecific parameters. Furthermore, there is not an assumption regarding a fixed structure of the model which means models can grow in size to explain the complexity of the data. The main application of the nonparametric methods is for the cases which the data is ranked, but there is not any clear numerical interpretation. They can be applied when the less is known and the more robust model is needed [15, 16].

The structure of the nonparametric model is not known prior and should determine from the data. It does not mean that there is not any parameter associated with the nonparametric model. Indeed, the parameters are more flexible and not fixed prior [17]. Figure 6.4 presents a few tests for each of the parametric and nonparametric approaches. Below are a few examples of the nonparametric models [18, 19].

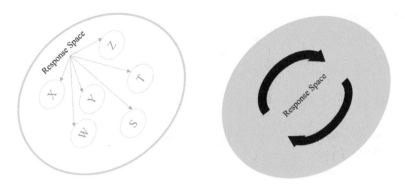

Fig. 6.5 Response space for categorical (left figure) and numerical (right figure) variables

- A simple "histogram"
- Kernel density estimation
- Nonparametric regression
- K-nearest network
- Support vector machine

6.2 Data Requirement

Data requirements should be clearly defined in the early stages of the analysis based on the end users' needs. It should be always considered that the cost of the analysis exponentially grows as the size of the data increases [3]. Data analysts should first completely understand the customers' expectations and desired outputs. In the next step, variables should be selected based on the physics governing on each system or empirical findings. Some of these variables might end up being insignificant after the statistical analysis [20]. The purpose of this step is trying to understand how the system works as accurate as possible. Variables can generally categorize as numerical or categorical variables. Categorical variables can only take one value out of the limited options for the variable value, while the choices for numerical variable either in limited or unlimited space are unlimited. For instance, left response space of Fig. 6.5 might be a random categorical variable which indicates the tomorrow's weather [21]. The options for this variable are limited to a few choices such as rainy, sunny, cloudy, windy, etc. Right response space of the Fig. 6.5 can be defined as a random numerical variable which indicates the chance of having a rainy day. The outputs can be any values between 0 and 1 [22]. Figure 6.5 shows the response space for the numerical and categorical variables.

Table 6.1 Data types

Statistics	Programming language
Countable data	Integer
Binary data	Boolean
Real value data	Floating point
Categorical data	Enumerated
Random vector	Array
Random tree	Tree

6.3   Data Type

Data can have various types based on each way of definition, application, and implementation. Various programming languages might use different terminologies, but most of them are sharing the same definition for the data type as Table 6.1 shows [23, 24].

6.4   Data Process

Data process refers to a set of activities which prepares the collected raw data as an input for the next steps of analyses. Collected raw data should be organized and placed into the columns and rows based on the standard of each application with respect to the developed algorithms and models. As an example, it might be possible that various sensors have been installed to measure the same type of pressure for different units. Each sensor manufacturers probably used some different characters for each attribute. What is important here is that all the collected measurements should be processed based on the same procedure to be ready to inject to the algorithms and models. One of the reasons is that the algorithms are usually developed for general application, and it would be tough to tailor them for each system [25, 26].

6.4.1   Data Collection

The role of the data for the analysis is similar to the role of the blood for a human body. Reliable outcomes cannot be generated without feeding the well-preprocessed raw data into the algorithms and models. Test data sets can be used to develop models or algorithms, but eventually, the collected data should be considered as the real inputs to obtain valid and reliable results. Data may be collected from the sensor measurements or online sources. For instance, the ambient temperature of a region might be collected using the satellite outputs instead of sensor measurements. This might be helpful to transfer less amount of data through each step of the analysis. Furthermore, online resources usually provide the filtered data which can be directly fed into the algorithms and models [27, 28].

6.4.2   Data Preparation

Data preparation activities involve the efforts to prepare or preprocess the raw data or data sources in order to refine the most beneficial information. The outcomes of the data preparation techniques can highly affect the efficiency and performance of the downstream businesses which mostly perform the data analytics. Data preparation is one of the most important initial actions in the data analysis stream. The efficient techniques can enfold the data content and make them more available and accessible by other users. Data preparation efforts can be categorized in various discrete processes. A few of these processes are data fusion, data cleansing, data augmentation, and data delivery [29–31].

6.4.2.1   Data Fusion

Data fusion refers to integrating the data coming from various resources to make them more consistent and accurate. The output of the data fusion techniques should provide more information compared with any of the individual resources. Data fusion efforts can be classified as low, intermediate, and high based on the stage of the process which these techniques take place. The low-level data fusion refers to the raw data integration to produce a richer raw data set. For instance, the outdoor air temperature might be recorded by various sensors in approximately the same area. Data fusion techniques can be performed on these set of measurements to integrate the raw information for generating a more accurate and reliable data set [32].

6.4.2.2   Data Cleansing

After the initial processes on data, it might still contain incomplete, duplicate, or error data. Data cleansing is a process of preventing or correcting any of these noises. These noises might cause some inconvenience in the analysis which will be performed in the next steps. One of the actions of data cleansing is related to matching or synchronizing the data. Each of the rows of the attributes should share the same time stamp. It means that if the analyst decides to eliminate one of the rows of one attribute, the same row with the same time stamp should be taken out for all the attributes. In the end, before feeding the data into the algorithms, the dimensions of all the attributes should be the same [33, 34].

6.4.2.3   Data Wrangling

Data wrangling or data munging refers to a set of data transformation, and data mapping activities perform on one set of raw data form to convert it to another format [35]. This process makes the data set more valuable and appropriate for a variety of downstream businesses. This includes mugging, data visualization, data aggregation, training data sets, etc. Data munging usually denotes the process of data transformation with the aim of enhancing the users and customer understanding of the data. For instance, data punctuations, data removing, and HTML tags are a few examples of data munging. All these methods start from preparing the raw data for feeding to the predefined data structure and end with depositing the results in the data lakes or data storages [36, 37].

6.4.2.4   Data Scraping

Sometimes the output of one computer program is considered as the input for another one. The process of extracting the data from the output of one computer program is called data scraping. Most of the time, there are well-defined protocols and standards for each computer program which make the data transformation easier with minimum ambiguity and mostly automated through pieces of structured codes. Most of the time, these data transformations are not human readable. Bar codes are an example of transformed data which are not easily readable by a human. Data scraping is considered as an inelegant ad hoc method which is sometimes called the last resort when no other technique is available or applicable [38].

6.4.2.5   Data Filtration

Data filtering refers to a wide range of activities which all should be performed to refine the data set to what the users need. These techniques try to take out the portions of the data which are repetitive, redundant, empty, noisy, irrelevant, and sometimes very sensitive. Some other filtration methods might limit the accessibility of the data set for each type of users. For instance, the Social Security number should be set a private data set which means everyone cannot have access to this information. Data filtering techniques try to reduce the content of the noise or error which inherently might be attached to the collected raw data [39, 40].

6.4.2.6   Data Mapping

The process of creating the connections between the elements of two or more different sources of data is called data mapping. Data mapping techniques can be applied to detect any hidden relationship or pattern between the data sources. The

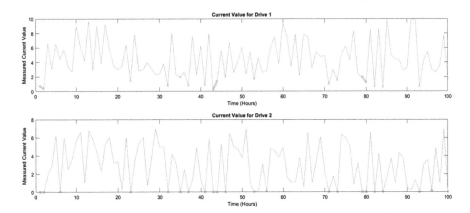

Fig. 6.6 Example of data mapping between two sources of current

newest version of the data mapping techniques is based on the data-driven algorithms which involve the simulations data evaluation of the data sources. These
evaluations are mostly based on the heuristic or statically algorithms which are able
to discover the complex and complicated mapping between two data sources. For
instance, Fig. 6.6 presents the current value for two different electrical drives given
the same sampling number in the same interval of time. Mapping the data coming
from these two sources can reveal the fact that anytime that the drive 1 is working in
low power (current under 2 Ampere), drive 2 is shut down (currently equal to zero)
[41, 42].

6.5 Data Visualization

Data visualization or visual communication is one of the most important phases of
the data analysis to visually summarize the output of the algorithms, models, and
processes. Data visualization techniques seek to graphically represent the information and data. Indeed, data visualization activities involve the study of the visual
representation of the data in both science and art areas to encode the data as a
graphical object. The outputs can be presented in the form of statistical graphics,
plots, figures, bars, and other tools [43]. An efficient visualization can give this
opportunity to the users to understand the data behaviors, the reason for further
analysis, and the extracted output of the raw data. One of the advantages of this step
is making the data more understandable, accessible, and usable. Data visualization
can be applied in each phase of the analysis starting from the raw data collection to
the finalized report presentation. Consequently, various tools might apply based on
each application. For instance, the visualization tools might be used to compare
several attributes or perform the casualty analysis. According to Friedman (2008),
the "main goal of data visualization is to communicate information clearly and

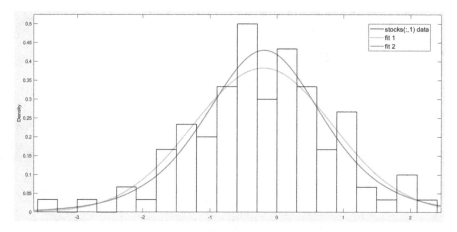

Fig. 6.7 An example of a simple dashboard generated by MATLAB software

effectively through graphical means. It doesn't mean that data visualization needs to look boring to be functional or extremely sophisticated to look beautiful" [44]. Data visualization is a key tool in the age of the big data, which trillions of rows of data may be generated in every day. Visualization techniques tell the story regarding the raw data in a form easier to understand. One of the main purposes of data visualization techniques is to highlight the trend, outlier, and abnormalities to quickly obtain the analysts' attention. It should be noted that the balance between the form and functions is very important in order to present the most beneficial information. Data presentation should be stunning enough to catch the attention of the analysts to convey the right message. Therefore, data visualization tools try to make the data more understandable. Figure 6.7 presents an example of a simple dashboard generated by MATLAB. The following states a few important features of an efficient data visualization tool:

- Fast analytics which quickly connect and visualize the data
- Big data ranging from Excel spreadsheet to the trillions of rows stored on cloud-based services such as Hadoop or Microsoft Azure
- Automatic update to refresh the data based on the defined scheduled
- Easily usable for majority of the users who do not have deep inside into programming
- Efficient dashboard design to get the greatest insight by using several views of the data

It should be considered that data visualization tools are different for classical and modern applications. As mentioned earlier in this chapter, the main reason is the size of the data. With the increasing applications of the Internet and Internet of Things (IoT) devices, the concept of the big data analysis has earned more attention. As the size of the data increases, a more preprocessing analysis should be performed on the raw data. For instance, assume that a smart device is constantly measuring the

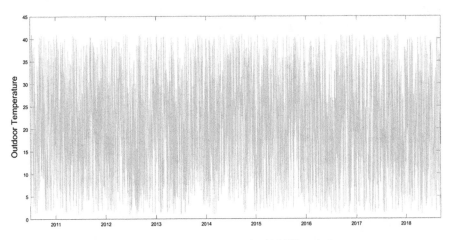

Fig. 6.8 Collected raw data from temperature sensor for 100,000 periods

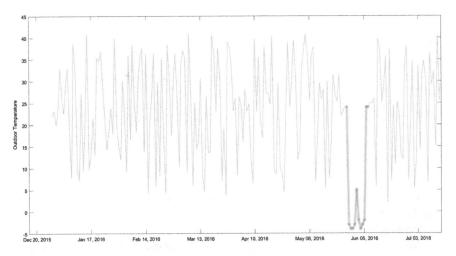

Fig. 6.9 Processed data for temperature sensor measurement

current of an electric drive two times every second. In order to analyze the monthly performance of the drive, most probably the visualization of the collected raw data over a month cannot reveal any beneficial information. The reason could be the enormous number of plotted data points. Figure 6.8 is an example of the data visualization for a collected raw data of an attribute coming from a smart sensor over 100,000 periods. There are a variety of techniques which can be performed to process the raw data to reveal more beneficial information. Figure 6.9 presents the data for the attribute after applying a simple technique such as averaging over a few periods [45, 46]. As can be seen, the user can easily notice an abnormal situation due to the rational visualization of the data.

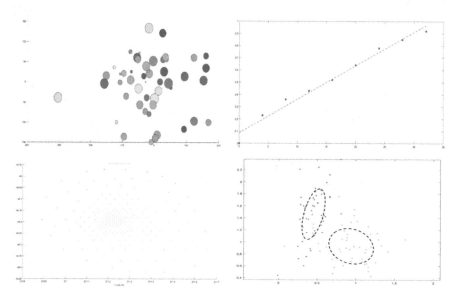

Fig. 6.10 A few examples of data visualization tools obtained by MATLAB software

Figure 6.10 presents a few examples of data visualization tools obtained by MATLAB.

6.6 Data Acquisition

Data acquisition starts from sampling the signals which measure the real-world physical condition of the system under the study. In the next step, the collected signals should be prepared for feeding to the algorithms installed on either on-site or remote computers for further analysis. All the collected measurements should be converted to a digital numeric value which the computers and programming languages can manipulate them. All these tools try to acquire data from various hardware equipment. Furthermore, these actions attempt to unify the collected data and signals as much as possible. Most of the time, there are some standards and protocols which the developed applications rely on to unify the data coming from various hardware and software measurements. Data acquisition applications are usually controlled by software or programming language [47]. Figure 6.10 depicts a big picture of a digital data acquisition system. The main components of the data acquisition are as follow:

- *Sensors*: Collect the real-time measurements of the physical conditions of the units as the electrical signals.
- *Signal Conditioners*: Process the collected raw signals to make them ready in a form which can be converted to a numeric digital value.

- *Signal Converters*: Convert the collected signals into the numeric values which can be directly fed into the computers.

In general, the waveforms can be classified as periodic and aperiodic. Periodic waveforms are generated based on the regular interval which presents the same periodic wave. Aperiodic waveforms might show the same shape of the wave but in a changing interval. Figures 6.11 and 6.12 present examples of periodic and aperiodic waveforms. The other case of the aperiodic waveforms is when the shape of the signal is not following the same pattern [48, 49] (Fig. 6.13).

6.7   Data Lake

Data lake is a repository in which each organization keeps its data in a natural format. A data lake is usually a single place where all the entities of an organization put the raw copy of the system data and analyzed the data for future uses. Data lake might include the structured data consisting of columns and rows, semi-structured data (e.g., CSV, XML), unstructured data (e.g., PDF), and binary data (e.g., audio). There are various companies who offer the storage services. The most commonly used examples are Microsoft Azure, Apache Hadoop, and Amazon S3. Although there are various advantages of data lake storage services, there are some criticisms as well. As Sean Martin, CTO of Cambridge Semantics stated, "We see customers creating big data graveyards, dumping everything into HDFS (Hadoop Distributed File System) and hoping to do something with it down the road. But then they just lose track of what's there. The main challenge is not creating a data lake but taking advantage of the opportunities it presents" [50]. The most successful companies are constantly evaluating their data lake to figure out which data or metadata is important to their business [51].

6.8   Dark Data

A portion of a data set is named as the dark data if it passes through the initial analysis but cannot provide any helpful information to be used in the decision-making process. An efficient data analytic approach can highly enhance the capability of an entity without relying on the size of the collected data. This indicates that one organization might analyze its available data more efficiently although less data is being collected compared with other competitors. This feature can bring business advantages to an entity. It should be always considered that as soon as data is being recorded, an associated cost has emerged. The cost can include collecting, streaming, transferring, and storing costs. Therefore, the dark data exists if the same level of insight can be reached by collecting less amount of data [52]. It means that a part of the data remains underutilized while the organization already spend money to have it

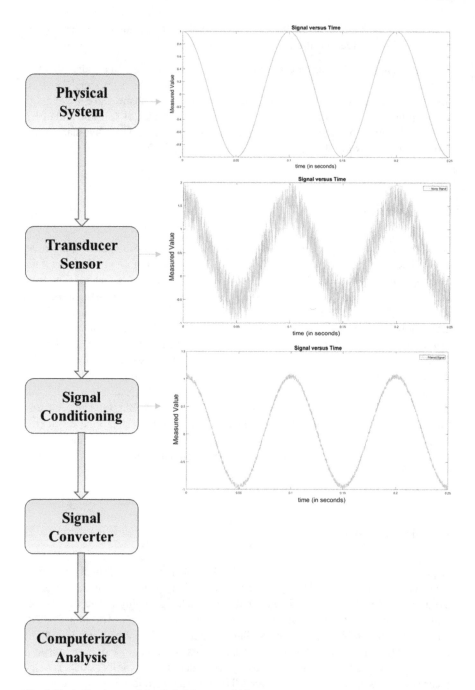

Fig. 6.11 A big picture of the digital data acquisition system

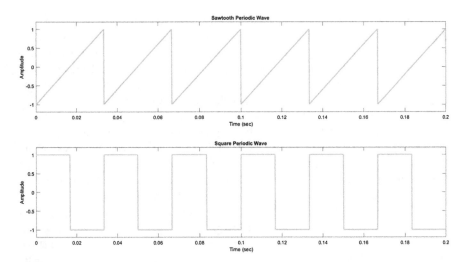

Fig. 6.12 Examples of periodic waveforms

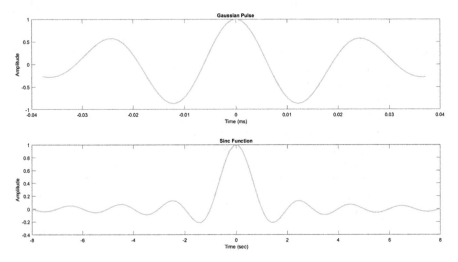

Fig. 6.13 Examples of aperiodic waveforms

available. A large portion of the dark data is unstructured which means the format of the information in the data is difficult to categorize or read by the computer to analyze. According to Computer Weekly, 60 percent of organizations believe that their own business intelligence reporting capability is "inadequate," and 65 percent say that they have "somewhat disorganized content management approaches" [53]. There are a lot of companies who are currently trying to develop a cognitive computer system which has the ability to analyze the unstructured data and extract some information out of the dark data sets.

As IBM estimated, roughly 90 percent of the data which has been collected with the smart sensors and analog to digital converters never get used. Some businesses believe that dark data is helpful in the future for reaching more information and business intelligence regarding how they performed during the past periods. This might be still applicable in many businesses due to the small cost of storage. On the other hand, data security remains challenging for most of the organizations.

There is a chance that a data set becomes a dark data set over time. This is called the "perishable data" after the "live flowing data" eras. According to IBM, about 60 percent of data loses its value immediately [54]. If the data analysis is being postponed, there is a higher chance that the data loses its value. For instance, in the field of fraud detection, if the issuer company does not analyze the transactions immediately, it might be too late to detect an abnormal transaction.

According to *The New York Times*, 90 percent of the energy used by data centers is wasted [55]. The associated costs of that portion of the data could be saved if it never was collected, streamed, transferred, and stored. According to Datamation, "the storage environments of EMEA organizations consist of 54 percent dark data, 32 percent Redundant, Obsolete and Trivial data and 14 percent business-critical data. By 2020, this can add up to $891 billion in storage and management costs that can otherwise be avoided" [56]. Continuous storage of the data can put an organization at risk especially in the case of breach. This can end up to serious repercussions and might highly affect the reputation of a business. For instance, the simplest example is related to the sensitive data for the research and development department of an organization. Unsecure data might let other competitors to access their high-level achievements.

The current era is called as a data-driven one which most of the objects surrounding us are somehow dealing with the data. Businesses should be able to manage the data flow through their data analytic approaches. Data and analytics are the two main basics of the modern industry, and there is no doubt that the prosperous organizations are somehow linked to the data and data-driven models. But, it should be always considered that data, information, algorithms, models, and achievements need precise approaches to keep them fresh and safe [57].

6.9 Big Data

The definition of the big data highly depends on the application, but, in general, a data set can be called as big if the traditional algorithms and models have difficulties to process and analyze the data. It could be due to either size or complexity of the data. Analysis of big data is usually beyond the ability of the most commonly used software to hand and generate a reliable result within a tolerable elapsed time [58]. The concept of the big data has got increasing attention during the last few decades as the application of the smart devices increased and, consequently, more data became available for analysis. A few examples of the big data challenges are data collection, transformation, storage, analysis, visualization, etc. Data sets grew

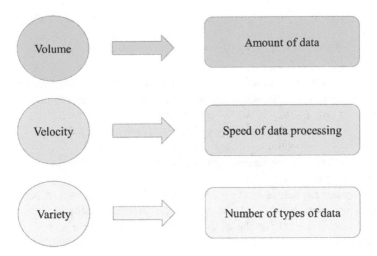

Fig. 6.14 3Vs feature of big data

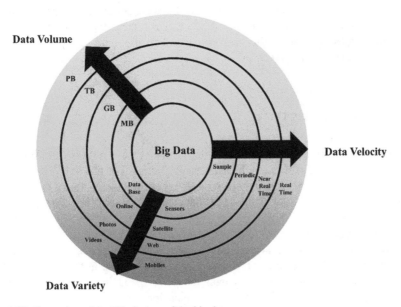

Fig. 6.15 Expansion of the 3Vs feature of the big data

exponentially due to the vast application of the Internet of Things (IoT) devices in a variety of applications [59]. For instance, each smartphone can constantly collect the data regarding user activities. It has been reported that the world's technological per capita capacity to store information has crudely doubled every 40 months since the 1980s. As Figs. 6.14 and 6.15 present, the big data originally had three main features. Another feature for the big data which can be added to the 3Vs is veracity.

Veracity concerns about the data quality to generate accurate results. Currently, the focus of the big data is not only on the size of the data. It mostly refers to the predictive algorithms and models which extract the most beneficial information out of the collected data [60].

Big data and the Internet of Things (IoT) interact with each other. Kevin Ashton, a digital innovation expert who is credited with coining the term, defines the Internet of Things in this quote: "If we had computers that knew everything there was to know about things—using data they gathered without any help from us—we would be able to track and count everything, and greatly reduce waste, loss, and cost. We would know when things needed replacing, repairing or recalling, and whether they were fresh or past their best" [61].

Data sampling is helpful to save time and cost. The analysts select a sample which is accurately representative of the system to drive a reliable conclusion. Data sampling techniques enable the users to estimate the characteristics of the whole population by using the right set of data points in a shorter period of time with less cost. For instance, suppose that the purpose of the analysis is monitoring the coil temperature of a drive. Temperature sensors might be able to measure and collect the temperature a few times every second. Indeed, the same outcome might be reached by using a sample data point every minute. It depends on the application to either sample or does not sample the whole data set but, it should always consider that larger data sets need more time and cost [62].

6.10 Time-Series Data

A series of data points indexed in time order is called time-series data set. Most of the time, the events are recorded in the same spaced interval of times. For instance, the time span for a smart device might be set as every second which means one record of measurement every second. Time-series data are widely used in economics, statistics, pattern recognition, signal processing, mathematics, finance, control engineering, forecasting, communication, etc. Time-series analysis refers to a series of activities which try to extract the beneficial information of the data points with time indexes [63]. As a result, time plays an important role in the analysis. As an example, consider the time-series data for wind speed in order to forecast the generation of a wind turbine. As a one-dimension array, wind speeds might not reveal useful information regarding the wind generation prediction. With the help of the time indexes, the analyst is able to understand the system behavior using the historically collected data to predict future outcomes. Time-series data for wind speed might divulge that the wind generation in a specific time of a day, such as midnight, might reach to its peak given specific operational and environmental conditions [64].

Time-series comparison is one of the most beneficial analyses of time-series data. Especially with an increasing trend of the IoT smart devices application and more available time-series data set, this tool has got more attention during the last decade.

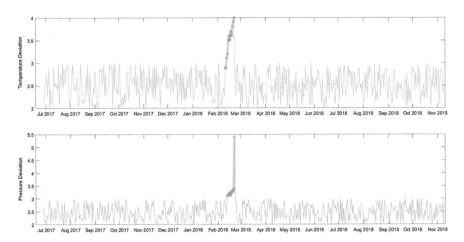

Fig. 6.16   An example of time-series data

More available data points can help the analyst to understand any possible relation between the system features. The overall trend or behavior of the features can be compared at each point of the time [65]. As an example, consider a high-voltage transformer as a component of the electrical distribution system. Suppose that the current maintenance policy is based on the "run to failure" which indicates that failure data set can be obtained over a few years of analysis. If the data analyst has the time series of the most statistically significant variables, some important information can be extracted. For instance, one of the conclusions can be regarding the abnormal efficiency of the unit a few periods before the failure. Excessive temperature, current, and voltage fluctuation might be the other possible outcomes. Therefore, as a result of the analysis, it might be possible to prevent a real-time preventive maintenance scheme to predict any upcoming failure a few periods in advance. In real-world application, early warning for failure detection is one of the most interesting and challenging topics. Statistical analysis of the time-series data generally reflects the fact that the data points which are closer together or happened on the same time span are more related together than those points which are falling apart. For instance, in weather forecasting application, the same time of the year over the last few years might disclose more information than the analysis of the last few days. Figure 6.16 represents an example of time-series data for two features. Suppose that a failure has happened during February to March 2018. The analysis of time-series data for various features might give an insight into an event. In this case, it can be considered that a temperature and pressure deviation behave abnormally (an increasing trend) a few periods before the failure. Algorithms and models can be retrained and learned from such past events to be smart enough to issue an early notice in similar cases. It should be considered that a conclusion such as the above example cannot be easily always reached. It probably needs more advanced analysis than visual.

Time-series methods can be classified as time-domain and frequency-domain. In time-domain, the system analysis is based on the time of the data by presenting how the system changes over time. Frequency-domain analysis depends on the frequency rather than time and investigates the portion of the data or frequency band over a range of frequency. As an example, in the field of control engineering, a signal or mathematical function can be analyzed with respect to the frequency instead of time. Time-domain and frequency-domain can be converted by applying a series of functions called transform function. As an example, Fourier transform function converts the time-domain into the frequency-domain by sum or integral of the sine wave for different frequencies.

6.11   Data Mining

Data mining refers to the analysis of the large data sets with the purpose of discovering patterns and extracting the knowledge. Most of the data mining techniques have overlap with either machine learning (ML) or statistical methods. Data mining is one of the interdisciplinary areas of computer science which seeks to extract the information out of large data sets in an intelligent way. Data mining can be also called as a knowledge discovery method [66]. Data mining involves a variety of techniques, such as the techniques that have been discussed earlier in this chapter, and starts from the raw data collection to data analysis. The applications of data mining techniques are very broad. A few examples of these applications are data collection, extraction, storage, analysis, as well as any application including artificial intelligence (AI) methods [59]. It should be considered that the analyses are either fully automated or semiautomated due to the large sets of data. Figure 6.17 briefly describes various categories of data mining methods. The detail of these algorithms will be discussed in the next chapters of this book.

Data mining techniques can be classified into six main categories as follows:

1. Anomaly detection
2. Regression
3. Clustering
4. Classification
5. Association rule learning
6. Summarization

One of the main steps of the data mining process is data validation. It should be always kept in mind that the outputs of the software should not be misused. It means that it might be possible that the results of a sample under analysis seem significant, but the results should not be overgeneralized to other similar populations without conservation. Therefore, the results of the analyses should be carefully reviewed before making any overall conclusion. Overfitting issue in machine learning (ML) algorithms is a simple example of validation need. It might be possible the data mining methods find a pattern in the data set which does not exist in the population or real system. A part of the data which has been selected for further

Fig. 6.17 Categories of
data mining

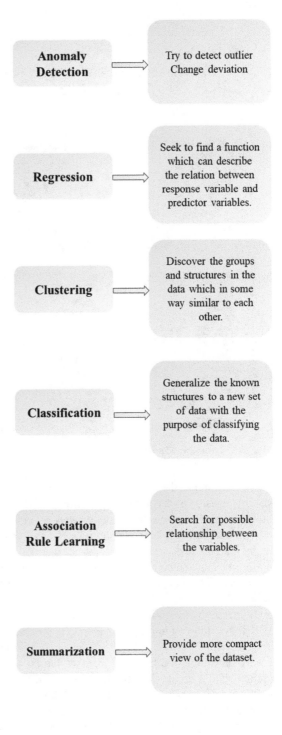

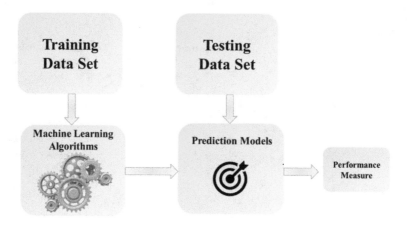

Fig. 6.18 A big picture of the model performance measurement procedure

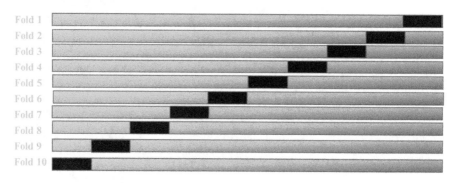

Fig. 6.19 Tenfold validation procedure

analysis is called the training data set. The results of the algorithms and models should be tested on the test data or verification data set. Test data is a data set which has not been used for training the models. It should always be considered that the training and testing data sets should not have any overlap. The overfitting issue might emerge, and the results might not be valid since the model performance has not been evaluated correctly. Figure 6.18 depicts a big picture of the model performance measurement procedure and the relation between the training and testing data sets.

If the size of the available data is small, the analyst can use other techniques such as K-fold validation in order to generate the training and testing data set which ends up a reliable measurement of the model performance. For instance, in K-fold validation techniques, the entire data should be divided into K equal sections. K-1 section should be used for training and the remaining part for the validation. The same procedure takes place K times, and the model performance measurements can be summarized as an average of all K measurements. Figure 6.19 presents a tenfold validation procedure. The detail of the evaluation statistics will be discussed in the next chapters of this book.

6.12 Concluding Remarks

Wide applications of the Internet and the latest developments of the Internet of Things (IoT) products have connected millions of devices together through the secured platforms. The current era can be called as the data-driven epoch since most of the objects surrounding us are in one way or another producing, receiving, or depending on the data. There is no doubt that the recent developments in fields of IoT, artificial intelligence (AI), Industry 4.0, machine learning (ML), etc. could not be successfully possible without the aim of data. This chapter explained the most important features of the data, types of the data, data preprocessing, data analytics, and step-by-step activities which are needed to extract the most beneficial information out of the raw data. Efficient utilization of the data brings the business advantages to the organizations which have more available data. Different types of cost are associated with a data set to become available as an input of the data analytics. Therefore, a business cannot be successful without the efficient utilization of the data. The fundamental activities for preprocessing the data to be prepared for downstream businesses have been discussed in this chapter.

References

1. C. Judd, G. McCleland, *Data Analysis* (Harcourt Brace Jovanovich, 1989). ISBN 0-15-516765-0
2. Data Cleaning. Microsoft research. Retrieved 26 Oct 2013
3. J.S. Bendat, A.G. Piersol, Random data analysis and measurement procedures. Meas. Sci. Technol **11**, 1825 (2000)
4. ConTaaS: An approach to internet-scale contextualisation for developing efficient Internet of Things Applications. ScholarSpace. HICSS50. Retrieved May 24, 2017
5. R.H. Shumway, Statistics and data analysis in geology. Technometrics **29**:4, 492–492 (1987)
6. H.R. Bernard, Qualitative data, quantitative analysis. CAM J **8**(1), 9–11 (1996)
7. H.R. Bernard, *Research methods in anthropology: Qualitative and quantitative approaches.* (Rowman & Littlefield, 2017)
8. J. Brannen, *Combining Qualitative and Quantitative Approaches: An Overview. Mixing Methods: Qualitative and Quantitative Research* (Routledge, London, 2017), pp. 3–37
9. C.C. Ragin, *The Comparative Method. Moving Beyond Qualitative and Quantitative Strategies With a New Introduction* (University of California Press, 2014)
10. W. Baker, Qualitative and quantitative analysis, in *The Creative Enterprise of Mathematics Teaching Research*, (Sense Publishers, Rotterdam, 2016), pp. 171–178
11. P. Mayring, T. Fenzl, Qualitative content analysis program qcamap–an open access text analysis software, 2016
12. Feng Xu et al., Qualitative and quantitative analysis of lignocellulosic biomass using infrared techniques: A mini-review. Appl. Energy **104**, 801–809 (2013)
13. S. Geisser, W.O. Johnson, *Modes of Parametric Statistical Inference*, vol 529 (John Wiley & Sons, Hoboken, 2006)
14. D.R. Cox, *Principles of Statistical Inference* (Cambridge University Press, 2006)
15. K.P. Murphy, Machine Learning: A Probabilistic Perspective (MIT press, Cambridge, MA, 2012)
16. D.A. Freedman, *Statistical Models: Theory and Practice* (Cambridge University Press, New York, 2009)

17. B.W. Silverman, *Density Estimation for Statistics and Data Analysis* (Routledge, Boca Raton, 2018)
18. J. Murray, Likert data: What to use, parametric or non-parametric? Int. J. Bus. Soc. Sci. **4** (11) (2013)
19. A.J. Vickers, Parametric versus non-parametric statistics in the analysis of randomized trials with non-normally distributed data. BMC Med. Res. Methodol. **5**(1), 35 (2005)
20. F. Mosteller, J.W. Tukey, *Data Analysis and Regression: a Second Course in Statistics. Addison-Wesley Series in Behavioral Science: Quantitative Methods* (Addison-Wesley, Reading, 1977)
21. J.W. Tukey, *Exploratory Data Analysis*, **2**, 131–160 (1977)
22. C. Chatfield, *Introduction to Multivariate Analysis* (Routledge, Boca Raton, 2018)
23. J.M. Linacre, Detecting multidimensionality: Which residual data-type works best? J. Outcome Meas. **2**, 266–283 (1998)
24. J.W. Thatcher, E.G. Wagner, J.B. Wright, Data type specification: Parameterization and the power of specification techniques. Proceedings of the tenth annual ACM symposium on theory of computing. ACM, 1978
25. A. Langley, Strategies for theorizing from process data. Acad. Manag. Rev. **24**(4), 691–710 (1999)
26. M.C.H. McKubre et al., Measuring techniques and data analysis, in *Impedance Spectroscopy Theory, Experiment and Applications*, (2018), pp. 107–174
27. L.A. Palinkas et al., Purposeful sampling for qualitative data collection and analysis in mixed method implementation research. Admin. Pol. Ment. Health **42**(5), 533–544 (2015)
28. M. Cleary, J. Horsfall, M. Hayter, Data collection and sampling in qualitative research: Does size matter? J. Adv. Nurs. **70**(3), 473–475 (2014)
29. J. Patel et al., Predicting stock and stock price index movement using trend deterministic data preparation and machine learning techniques. Expert Syst. Appl. **42**(1), 259–268 (2015)
30. K. Coussement, S. Lessmann, G. Verstraeten, A comparative analysis of data preparation algorithms for customer churn prediction: A case study in the telecommunication industry. Decis. Support. Syst. **95**, 27–36 (2017)
31. M.P. Panning et al., Verifying single-station seismic approaches using earth-based data: Preparation for data return from the InSight mission to Mars. Icarus **248**, 230–242 (2015)
32. M. E. Liggins, D. Hall, J. Llinas (eds.), *Handbook of Multisensor Data Fusion: Theory and Practice* (CRC Press, Boca Raton, 2017)
33. Z. Khayyat et al., Bigdansing: A system for big data cleansing. Proceedings of the 2015 ACM SIGMOD international conference on management of data. ACM, 2015
34. I. Gemp, G. Theocharous, M. Ghavamzadeh, Automated data cleansing through meta-learning. AAAI, 2017
35. T. Rattenbury et al., *Principles of Data Wrangling: Practical Techniques for Data Preparation* (O'Reilly Media, Inc, Sebastopol, 2017)
36. I. G. Terrizzano et al., Data wrangling: The challenging journey from the wild to the lake. CIDR, 2015
37. T. Furche et al., Data wrangling for Big Data: Challenges and opportunities. EDBT, 2016
38. J.K. Hirschey, Symbiotic relationships: Pragmatic acceptance of data scraping. Berkeley Technol. Law J **29**, 897 (2014)
39. R. Sood, S. Garg, P. Palta, A novel approach to data filtration against packet flooded attacks in cloud service. J. Netw. Commun. Emerg. Technol **6**(5), 37 (2016)
40. D. Khachane, A. Suryawanshi, D. Pathak, Data selection and filtration technique for tuning virtual sensor model of NOx estimation in MATLAB. Computing communication control and automation (ICCUBEA), 2016 International Conference on. IE, 2016
41. H. Yoon et al., Efficient data mapping and buffering techniques for multilevel cell phase-change memories. ACM Trans. Architect. Code Optim **11**(4), 40 (2015)
42. F. Mazzarella et al., Discovering vessel activities at sea using AIS data: Mapping of fishing footprints. Information fusion (fusion), 2014 17th international conference on. IEEE, 2014

43. K. Cline et al., SU-F-T-99: Data visualization from a treatment planning tracking system for radiation oncology. Med. Phys. **43**(6Part14), 3484–3484 (2016)

44. M.O. Ward, G. Grinstein, D. Keim, *Interactive Data Visualization: Foundations, Techniques, and Applications* (AK Peters/CRC Press, Boca Raton, 2015)

45. T. Sander et al., DataWarrior: An open-source program for chemistry aware data visualization and analysis. J. Chem. Inf. Model. **55**(2), 460–473 (2015)

46. C. Donalek et al., Immersive and collaborative data visualization using virtual reality platforms. Big Data (Big Data), 2014 IEEE international conference on. IEEE (2014)

47. D.N. Mastronarde, Advanced data acquisition from electron microscopes with SerialEM. Microsc. Microanal. **24**(S1), 864–865 (2018)

48. W. L. Starkebaum et al., Waveforms for electrical stimulation therapy. U.S. Patent No. 9,937,344, 10 April 2018

49. M.-Z. Poh et al., Diagnostic assessment of a deep learning system for detecting atrial fibrillation in pulse waveforms. Heart **104**, 1921 (2018). heartjnl-2018

50. B. Stein, A. Morrison, Data lakes and the promise of unsiloed data (pdf) (Report). Technology Forecast: Rethinking integration. PricewaterhouseCooper, 2014

51. N. Miloslavskaya, A. Tolstoy, Big data, fast data and data lake concepts. Proc. Comput. Sci **88**, 300–305 (2016)

52. Ce Zhang et al., Extracting databases from dark data with deepdive. Proceedings of the 2016 international conference on management of data, ACM, 2016

53. Dark data could halt big data's path to success. ComputerWeekly. Retrieved 2015-11-03

54. Digging up dark data: What puts IBM at the forefront of insight economy | #IBMinsight. SiliconANGLE. Retrieved 2015-11-03

55. J. Glanz, (2012-09-22). Data Centers waste vast amounts of energy, belying industry image. The New York Times. ISSN 0362-4331. Retrieved 2015-11-02

56. Enterprises are Hoarding 'Dark' Data: Veritas - Datamation. www.datamation.com. Retrieved 2015-11-04

57. Leveraging Dark Data: Q&A with Melissa McCormack - Predictive Analytics Times - predictive analytics & big data news. Predictive Analytics Times. Retrieved 2015-11-04

58. S. John Walker, *Big Data: A Revolution that will Transform how we Live, Work, and Think* (2014), pp. 181–183

59. Xindong Wu et al., Data mining with big data. IEEE Trans. Knowl. Data Eng. **26**(1), 97–107 (2014)

60. Min Chen, Mao Shiwen, Y. Liu, Big data: A survey. Mob. Netw. Appl **19**(2), 171–209 (2014)

61. M.D. Assunção et al., Big Data computing and clouds: Trends and future directions. J. Parallel Distrib. Comput. **79**, 3–15 (2015)

62. R. Kitchin, Big Data, new epistemologies and paradigm shifts. Big Data Soc. **1**(1), 2053951714528481 (2014)

63. E.C. e Silva et al., Time series data mining for energy prices forecasting: An application to real data, in *International Conference on Intelligent Systems Design and Applications*, (Springer, Cham, 2016)

64. Yan Zhu et al., Time series chains: A novel tool for time series data mining. IJCAI, 2018

65. F. Martínez-Álvarez et al., A survey on data mining techniques applied to electricity-related time series forecasting. Energies **8**(11), 13162–13193 (2015)

66. I.H. Witten, E. Frank, M.A. Hall, C.J. Pal, *Data Mining: Practical Machine Learning Tools and Techniques* (Morgan Kaufmann, San Francisco, 2016)

Chapter 7
Data Analytics

7.1 Data Analytics

Data analytics is the process of examining the data in order to draw a conclusion regarding the information they may contain. Data analytics have been widely used in both industry and academic areas. Industries seek to make more informed decisions, and researchers try to verify scientific models, hypotheses, and theories. Data analytics are so close to business analytics (BA) in terms of reporting and online analytical processing [1]. Analytics has the power to change what organizations are going to do at which points of the time. Data analytics can highly boost businesses performances by enhancing revenue, cost, resources utilization, and operational efficiency. Furthermore, data analytics enable businesses to quickly identify and respond to the changes in either market or demand. Depending on the application, the data which passes through analytical tool may include a historical record or real-time information [2].

Data analytics methodologies can include the explanatory data analysis, to find any relation or pattern in the data sets, or confirmatory data analysis to apply statistical tools for verifying hypotheses regarding the data sets. High penetration of smart devices would provide more available information, in terms of big data, regarding the system under study [3]. On the other hand, the power of the analytical tools, such as machine learning (ML) or artificial intelligence (AI) tools, could highly enhance the effectiveness of the data analytics. Data analytic tools may have access to the managers at various levels to keep track of the processes more conveniently [4].

© Springer Nature Switzerland AG 2020
F. Balali et al., *Data Intensive Industrial Asset Management*,
https://doi.org/10.1007/978-3-030-35930-0_7

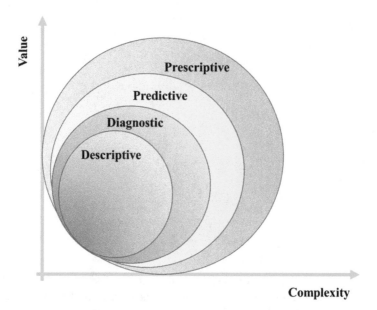

Fig. 7.1 Data analytics value versus complexity

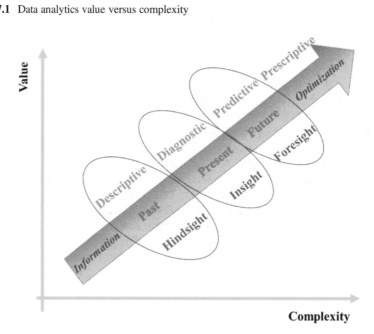

Fig. 7.2 Data analytics information, value, and complexity

7.2   Analytics Value

The main focus of this chapter is on the value of data analytics. Figures 7.1 and 7.2 present the value of each of the analytics versus the level of the complexity. The complexity of the analysis increases as moving from the descriptive toward the prescriptive analysis [5]. One of the reasons could be because of the level of uncertainty of the analytics [6]. Descriptive analysis has the lowest level of uncertainty since most of the factors are known. Predictive and prescriptive analysis have higher uncertainties since the data analytics are trying to provide useful information regarding the future [7].

Data analytics can have various applications depending on the time of the execution. In general, analytics can be classified as descriptive, diagnostic, predictive, and prescriptive. As Fig. 7.2 depicts, descriptive analysis is providing hindsight regarding the system behavior during the past periods [8]. The analyses of this step rely on historical recorded data. The analysts try to understand the system manners given various operating conditions. It should be considered that what happened in the past might never happen in the future. For instance, a system which experiences a permanent degradation process might never expose to the as good as new conditions.

Diagnostic analytics are mostly related to the current period of times or periods which are close to current. As an example, the control limits might indicate an abnormal situation. In this case, further investigations are needed in order to diagnose the system and restore it to the expected conditions [9]. Predictive analytics are trying to capture the future performance of the system with respect to the past and current working conditions. Predictive analytics provides a foresight regarding the system manners in the future. Each of the analytics will be discussed in detail [10].

Predictive analytics turn the data into beneficial information. Indeed, the predictive analytic tools try to extract information based on the observed data in order to predict the future outcomes or likelihood of an event occurring. Prescriptive analytics automatically synthesizes the big data, various sciences, business rules, and machine learning (ML) algorithms to suggest the decisions or set of actions for each prediction [11]. Figure 7.3 presents the four types of data analytics in more detail.

7.3   Descriptive Analytics

Descriptive analytics represent the key metrics and measures of an organization in order to demonstrate what had happened through various sectors of business. This type of analytics is probably the most common form of analytics, which has been widely used in various organizations. Monthly profit or lost report is one of the most tangible examples of descriptive analytics. Descriptive analytics try to look at the data representing the past to find a rational insight regarding the future status of the business [12]. Descriptive analytics tools reveal the reason behind the success or

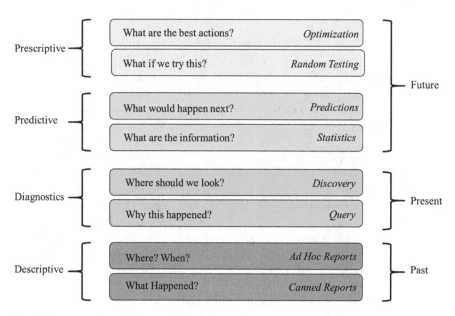

Fig. 7.3 An overview of four types of data analytics

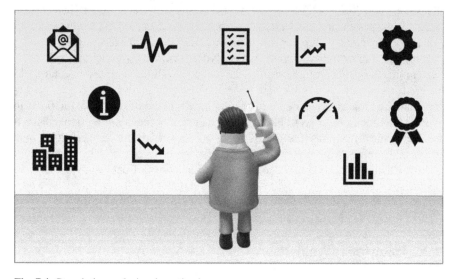

Fig. 7.4 Descriptive analysis schematic view

failure of events, which happened in the past, by mining the historical data. Almost all of the managerial boards are still using this type of data analytics (Fig. 7.4).

Descriptive analytics does exactly what the name implies. They seek to summarize the raw data and information in a way which are easily human interpretable. In other words, they describe the past. Descriptive analytics quantify the relationship

Fig. 7.5 Diagnostics analysis schematic view

between the explanatory variables, which are affecting the organization performance metrics and measures, with the main purpose of classifying the parameters or prospects into the groups.

7.4   Diagnostic Analytics

Diagnostic analytics empower the analysts to drill down and perform the root cause analysis in order to find the main causes of the events. Business information (BI) and well-designed dashboard incorporating the time-series data would enable the analysts to quickly find an overview of the organization standing point [13]. Indeed, this step is using more advanced tools to answer the question "why did it happen?". Data discovery, data mining, drill down, and correlation are the main characteristics of diagnostic analytics. Diagnostic analytics can provide a deep insight into a particular problem, which can be helpful for future events. For instance, the surprising value of the loss statement might motivate the analysts to request more detail report to reveal the reason. The detail report may indicate the employee working hours as one of the causes. Therefore, diagnostic analytics seek to always answer the questions regarding the reason for the events. Figure 7.5 depicts a schematic view of diagnostic analytics.

Fig. 7.6 Predictive analysis schematic view

7.5   Predictive Analytics

Predictive analytics is primarily about forecasting future conditions. It could be the likelihood of an event happening in the future, estimating a point of the time which an event might happen, or forecasting a quantifiable amount of an unknown variable by applying the developed predictive models. In general, predictive models utilize the data in order to make a prediction regarding the future condition of the same system. It should be noted that there is always an uncertainty attached to the predictive models, but they are still robust to enhance the decision-making process. The ability of the predictive models has been significantly improved as more data become available through the high penetration of the smart devices. On the other hand, the principles of the machine learning (ML) and the Internet of Things (IoT) could significantly empower the prediction process [14].

Predictive models seek to turn the raw data into a piece of valuable and actionable information. Predictive analytics deploy a variety of statistical tools such as data mining, machine learning (ML), and game theory to use the current and historical data to make the prediction regarding future outcomes. Indeed, predictive models are in charge of discovering the future risks and opportunities. Decision analysis and optimization are usually taking place within the predictive models. The overall objective of the predictive models can be summarized as obtaining an accurate and actionable insight regarding the future based on the observed data (Fig. 7.6).

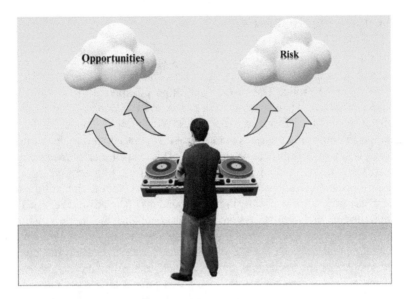

Fig. 7.7 Schematic view of prescriptive analytics

7.6 Prescriptive Analytics

Prescriptive analytics help the analysts to determine an optimal course of action based on the answers to the variety of the questions regarding "what might happen?". For instance, the analysts might have several options to dispatch the maintenance actions for a particular asset. The time-varying cost of maintenance actions needed items to be repaired or replaced, and the risk associated with each of these decisions can determine the optimal dispatch. Prescriptive analytics synthesize the big data, various principles of sciences, business rules, and IoT disciplines to take advantages of the predictions and take the most optimum decisions. Prescriptive analytics goes beyond the prediction. Indeed, in addition to the "what will happen" and "when will happen" questions, they should be able to justify the "why it would happen" questions (Fig. 7.7).

Prescriptive analytics are able to take advice on each of the outcomes of the prediction in order to help the businesses to take action effectively. The key roles of artificial intelligence (AI) and big data could have enhanced the effectiveness of prescriptive analytics during the last few decades.

7.7 Concluding Remarks

Various forms of analytics may provide a varying amount of value to a business. They all are important and have their own place. Data analytics seek to bring more values to an organization. The broad application of smart devices could highly enhance the effectiveness of data analytics. The size of the data is growing exponentially as the number of smart devices raises. What bring the business advantages to the organizations is not the size or amount of data. Indeed, the efficient utilization of the data is key for an entity to be successful. This chapter mainly focused on four different types of data analytics. Descriptive analytics try to explain the events, which have occurred in the past. Diagnostic analytics attempt to find out the reason for occurring each of these events. Predictive analytics take advantage of historical data and current status of the system in order to predict the future condition or likelihood of occurrence of an event. Prescriptive analytics make optimum decisions based on the prediction outcomes and their consequences. It should be considered that the complexity and uncertainty of the data analytics would increase as we move forward in prediction time. Descriptive and prescriptive analytics have the least and the most level of complexity and uncertainty, respectively.

Artificial intelligence (AI) and big data could highly affect the effectiveness of data analytics, especially where the uncertainty and complexity are considerable. Statistical analyses are usually trying to approve or reject a hypothesis by assessing the correlation among the data. Machine learning (ML) is mostly about predicting the possible outcomes in the future based on various variables. Businesses are able to bring more values to their organizations by taking advantages of the AI and ML tools for making a decision over time. It should be noted that there is usually a positive correlation between the business size and its needed data analytics. Descriptive and diagnostic analytics are more interesting for managers who usually take the low-level decision in terms of the time horizon. Predictive and prescriptive analytics are usually more interesting for managers who take high-level decisions for a longer horizon.

References

1. A. Gandomi, M. Haider, Beyond the hype: Big data concepts, methods, and analytics. Int. J. Inf. Manag. **35**(2), 137–144 (2015)
2. V. Grover, R.H. Chiang, T. Liang, D. Zhang, Creating strategic business value from big data analytics: A research framework. J. Manag. Inf. Syst. **35**(2), 388–423 (2018)
3. M.M. Najafabadi, F. Villanustre, T.M. Khoshgoftaar, N. Seliya, R. Wald, E. Muharemagic, Deep learning applications and challenges in big data analytics. J. Big Data **2**(1), 1 (2015)
4. K. Ousterhout, R. Rasti, S. Ratnasamy, S. Shenker, B. Chun, Making sense of performance in data analytics frameworks, pp. 293–307
5. R. Soltanpoor, T. Sellis, Prescriptive analytics for big data, pp. 245–256
6. B. Raja, J. Pamina, P. Madhavan, A.S. Kumar, Market behavior analysis using descriptive approach, Available at SSRN 3330017

7. L.P. Perera, Industrial IoT to predictive analytics: a reverse engineering approach from shipping (2017)
8. M. Ouahilal, M. El Mohajir, M. Chahhou, B.E. El Mohajir, A comparative study of predictive algorithms for business analytics and decision support systems: Finance as a case study, pp. 1–6
9. M.G. Kibria, K. Nguyen, G.P. Villardi, O. Zhao, K. Ishizu, F. Kojima, Big data analytics, machine learning, and artificial intelligence in next-generation wireless networks. IEEE Access **6**, 32328–32338 (2018)
10. K. Vassakis, E. Petrakis, I. Kopanakis, Big data analytics: applications, prospects and challenges, pp. 3–20
11. M.G. Kibria, K. Nguyen, G.P. Villardi, O. Zhao, K. Ishizu, F. Kojima, Big data analytics, machine learning and artificial intelligence in next-generation wireless networks, arXiv preprint arXiv:1711.10089
12. D.L. Olson, D. Wu, *Predictive Data Mining Models*. Springer Singapore (2017)
13. E. Siow, T. Tiropanis, W. Hall, Analytics for the Internet of Things: A survey. ACM Comput. Surv **51**(4), 74 (2018)
14. P. Grover, A.K. Kar, Big data analytics: A review on theoretical contributions and tools used in literature. Glob. J. Flex. Syst. Manag. **18**(3), 203–229 (2017)

Chapter 8
Machine Learning Principles

8.1 Introduction

As mentioned earlier in this book, the raw data might not be able to reveal any beneficial information without being processed. Indeed, raw data are only the input to the algorithms in order to perform specific tasks to obtain information regarding how the system under study performs. As Fig. 8.1 presents, the output of the algorithms is based on the raw data fed to the models. Therefore, algorithms are adding value to the raw data in order to extract the information. The first step of the analysis starts with acquiring the data [1]. Data can be acquired online as the output of the sensors or smart devices or offline as historical, simulated, or generated data sets. Raw data should be preprocessed in order to be prepared for algorithms. Data synchronization feature selection, data transformation, outlier detection, and noise removal are a few examples of the data preprocesses [2]. Predictive analytics mainly focus on the development and application of the algorithms in order to make a prediction. It should be considered that prediction is not necessarily a temporal prediction regarding what will happen in the future. Prediction principles can be applied to predict an unknown value of a parameter [3].

Condition indicators are extracting more beneficial information out of the collected raw data. In this step, the analyst should decide how to feed the preprocessed data into the algorithms. The main role of the ML algorithms is to develop predictive models by using the training and verifying data sets [4]. The output of the ML algorithms could be either a continuous or discrete variable. As the statistics present, machine learning (ML) algorithms have become among one of the most attractive subjects for the researchers during the last few decades. There are various studies which focus on either developing or enhancing the ML algorithms for various applications. It should be noted that the basics of the ML algorithms are based on statistical principles. From the statistical point of view, ML algorithms have been developed as the integration of the statistical algorithms with computer science and data mining philosophies. The following are the basic definitions of the ML term [5]:

© Springer Nature Switzerland AG 2020
F. Balali et al., *Data Intensive Industrial Asset Management*,
https://doi.org/10.1007/978-3-030-35930-0_8

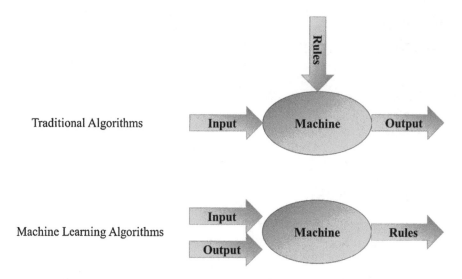

Fig. 8.1 Traditional versus ML algorithms overview

- *Machine*: A mechanically, electrically, or electronically unit which enables the operator to perform a task
- *Learning*: A process of gaining knowledge or skills regarding a system by any attempt such as studying, practicing, and experimenting

The main difference between the ML and traditional algorithms is in the way which the analyst defines the rules. As Fig. 8.1 presents, traditional algorithms are based on the predefined rules which are all coming from a series of logic which is able to extract the output with respect to the inputs. It means that there should not be any uncertainty or unknown entity for the traditional algorithms. It should be considered that more sophisticated rules are usually needed as the system becomes more complex. For this reason, it might be possible that the complex systems become unsustainable to maintain using the traditional algorithms. ML algorithms are supposed to overcome this issue since, in ML algorithms, machine is in charge of defining the rules between the inputs and outputs. ML programmer does not need to tailor the codes based on each new set of data. Indeed, the ML algorithms adapt to the new sets of information and evolve over time as exposed to more input data [6].

ML algorithms try to develop the models based on the training data sets. Feature vectors are extracted based on the training data sets, and ML algorithms are trying to understand more beneficial information regarding the system based on the training data set. The built model can be tested using the verification data set which does not have any overlap with the training sets. If the model acquires the acceptable performance measures, it can be applied for other cases to perform the prediction process. Figure 8.2 depicts an overview of the process starting from the training data sets to the accepted model which is able to make the prediction [7, 8].

The following is a summary of the ML steps (Fig. 8.3):

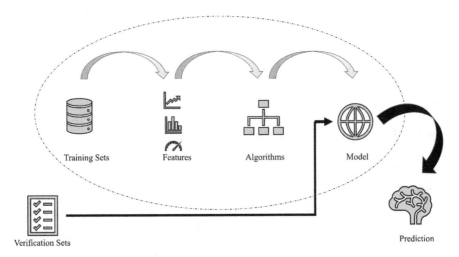

Fig. 8.2 An overview of the ML process

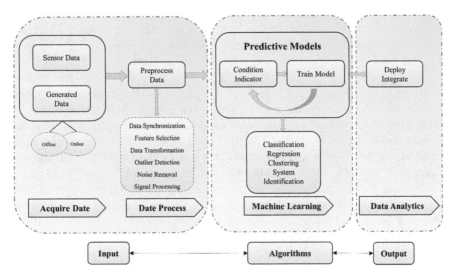

Fig. 8.3 An overview of the role of the ML algorithms among the data processes

1. Problem definition
2. Data collection
3. Data preprocess
4. Algorithms development
5. Algorithms performance evaluation
6. Feedback collection
7. Algorithms revisions
8. Algorithm selection

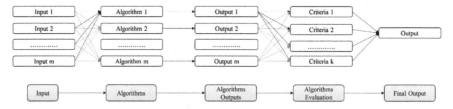

Fig. 8.4   An overview of the ML algorithms

9. Model development
10. Prediction

The main purpose of this chapter is to present the most commonly used statistical and ML algorithms more toward the application side rather than the theories behind the development of the algorithms. It should be considered that there are various books which only focus on one of the ML algorithms. Therefore, this chapter of the book is summarizing the most important applications and features of the ML algorithms. It should be noted that the term "machine learning (ML)" is not interchangeable with "artificial intelligence (AI)." Indeed, ML is a subfield of the AI which sometimes referred to as "predictive modeling" or "predictive algorithms." Created by an American computer scientist Arthur Samuel in 1959, the term "machine learning (ML)" is defined as the "computer's ability to learn without being explicitly programmed." As Fig. 8.1 presents, ML applies programmed algorithms the analyze the input data in order to predict an output value within an acceptable range. ML algorithms reach to more enhanced insight regarding the system under study over time as analyzing more data. Therefore, the performance of an algorithm is highly tied to the data feed into the machines. It means that the conclusions should be made given the data which the algorithms are exposed to. As the Fig. 8.4 presents, the general task for making a prediction regarding the future data points (Y) given the new inputs to the algorithms (Xs) is to find a form of mapping function (f) which is able to transform the inputs to the output. If the relation between the inputs and output is known prior, there is no need to apply the ML algorithms. In other words, the main role of the ML algorithms is to find the most applicable algorithm which is able to transfer the inputs into the outputs [9, 10]. There are various criteria to select the algorithm with the highest performance which will be discussed in detail in this chapter.

In the field of statistics and ML, what is being predicted is called the target variable. Data points are represented in terms of descriptive features. Pairs of features and target are called examples or instances. ML algorithms are automatically learning a model from the past data to make predictions regarding what would be the status of the target variable or what is the value of an unknown parameter. In other words, ML learns the relationship between features and target from historic examples, also called training data or training examples. Figure 8.5 depicts the relation between the features, target, algorithm, and predictive models. Model training and verification are the main two basic elements of the ML algorithms

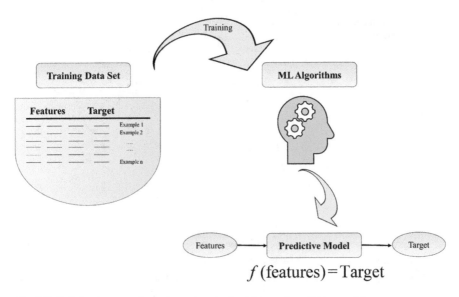

Fig. 8.5 Relation between the features, target, algorithm, and predictive models

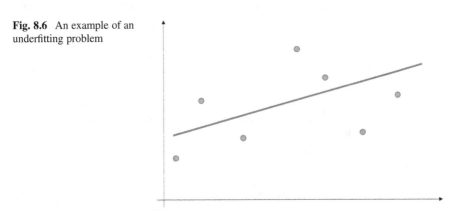

Fig. 8.6 An example of an underfitting problem

[11]. Predictive models try to learn from past examples using the testing data set. The performance of the algorithms can be verified using the verification data set which are unseen for the model during the training phase. It is very important to make sure that there is not any overlap between the testing and verifying data sets since the model performance can be highly affected [12].

It should be considered that ML algorithms are not supposed to do the magic. Indeed, they are supposed to explore and analyze the training data, which human brain is not able to handle that, to develop a predictive model which its performance is acceptable for the verification data set. Predictive models should be generalized which means performing satisfactorily for both training and verifying data sets. The following are showing the two overall kinds of mistake which can be observed as the results of the inductive bias of each algorithm (Figs. 8.6 and 8.7):

Fig. 8.7 An example of overfitting problem

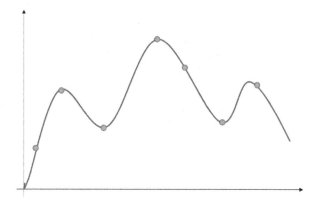

1. Underfitting occurs when the predictive model is too simplistic to represent the underlying relationship between features and target.
2. Overfitting occurs when the predictive model is too complex and fits the data too closely which result in enhancing the prediction noise.

One of the most famous theorems in ML area is called "no free lunch." This theorem indicates that there is not an algorithm which can globally perform better than others for all the applications. Based on this theorem, an analyst should have advanced knowledge regarding the ability of the algorithms in order to select the most applicable one [13]. For instance, it cannot be concluded that neural network (NN) is always performing better than decision trees or vice versa. Various parameters such as the size and structure of the data sets highly affect the performance of the algorithms. Consequently, various algorithms might be implemented for an application in order to reach the winner algorithm. ML algorithms are trying to study the data to obtain insight regarding the system under study. The applications of the ML algorithms are very broad and not limited to specific fields. Healthcare, energy, the stock market, fraud detection, image processing, signal processing, and weather forecasting are among the best examples of the fields which highly rely on the predictive models [13]. As Fig. 8.8 depicts, the main targets of the ML algorithms can be summarized as follows:

1. Categorizing the various classes (upper left)
   The value to be predicted is a nominal value such as positive or negative diagnosis.
2. Predicting a known outcome based on identified inputs (upper right)
   Most of the time, the value to be predicted is a numerical value such as stock prices and energy expenditures.
3. Identifying the patterns or specific relationships (lower left)
4. Detecting the abnormalities and unexpected behaviors (lower right)

ML algorithms can be classified based on four different categories. Before explaining each category, it should be noted that in general, there are two types of

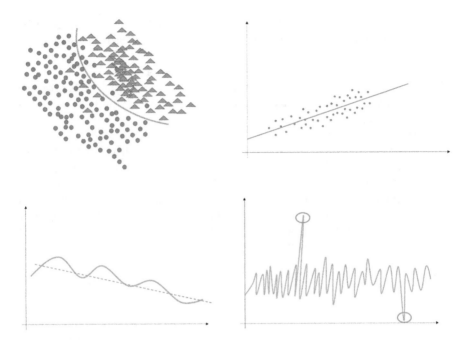

Fig. 8.8 The main targets of the ML algorithms

data as labeled or unlabeled data. Labeled data is a group of examples which are tagged with one or more labels.

1. *Supervised Learning*

In supervised learning, the algorithms are trained by rich examples of the anticipated inputs and outputs. Algorithms and models are going to use these examples to determine the correlations and logic that can be helpful to predict the target. Algorithms can be applied to other similar applications with some level of modifications once the predictive models are trained. Supervised learning algorithms are trying to develop predictive models based on the labeled training data set. The following are a few examples of the most commonly used supervised techniques [5]. A few of these methods will be discussed in detail in this book:

- Regression analysis
- Decision tree
- Support vector machine (SVM)
- Bayesian statistics
- Neural networks
- Random forest

2. *Semi-supervised Learning*

Semi-supervised learning algorithms are trying to address the problems as the same as supervised algorithms. The main difference between these two groups is in the training data set. As mentioned before, supervised learning algorithms are based on the labeled training data set. Semi-supervised learning algorithms can be applied when a part of the training data set remain unlabeled. In other words, some of the inputted are tagged based on the desired output, while the rest remain untagged. Semi-supervised algorithms are applicable for the cases which there are too much data to develop a comprehensive set of examples. In this case, the analyst let the algorithms to explore the general pattern or logic and then extrapolate the remaining untagged examples [14].

3. *Unsupervised Learning*

Indeed, unsupervised learning algorithms are trying to identify the patterns or relationships by parsing the available data. In this case, there is no answer sheet to develop the training example data sets. These types of algorithms are behaving close to the human brain. The intuition becomes more refine as the experiences grow. Anomaly detection and market basket analysis are among the examples of unsupervised learning algorithms [15]. The following are a few examples of unsupervised learning algorithms:

- Clustering
- K-mean clustering
- Nearest neighbor mapping
- Affinity analysis

4. *Reinforcement Learning*

Reinforcement learning is similar to teaching someone to play a game based on a series of predefined rules. In reinforcement algorithms, the set of available actions, rules, potentials, and states are provided to the algorithms. Algorithms are trying to obtain the desired outcomes based on the allowed rules and the reactions based on what algorithms have been explored. Algorithms are exploring various actions to maximize the reward. Indeed, algorithms should be able to select the best possible behavior or path in a specific situation [16, 17]. The following are a few examples of reinforcement learning algorithms:

- Artificial neural network
- Q-learning
- Markov decision process (MDP)

8.2   Machine Learning Algorithms

There are various ML algorithms which each might be applicable to specific applications. It should be noted that the detail of each of these algorithms might need a separate book. For instance, there are various available books which only explain the linear regression neural network, etc. The main purpose of this section is to summarize the application of the most commonly used supervised algorithms by presenting a brief review of the math behind each algorithm, overall objectives, main applications, and examples (Table 8.1).

Table 8.1 A few commonly used algorithms and general types

Algorithm	Type
Linear regression	Regression
Logistic regression	Classification
Decision tree	Regression/classification
Support vector machine	Classification
Naïve Bayes	Regression/classification
Neural network	Regression/classification

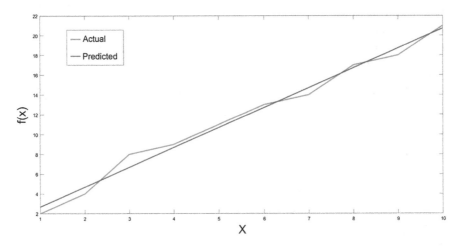

Fig. 8.9 An example of linear regression

8.2.1   Linear Regression

Linear regression is one of the most commonly used algorithms in the field of statistics and ML. The linear approach is a simple method for supervised learning. The main purpose of the algorithm is to find the numeric target value for the response variable given a set of independent variables, predictors, or features [18]. The response variable can be considered as a dependent variable. Figure 8.9 depicts a simple example of linear regression. It should be noted that the predicted function provided by the linear regression algorithm is always a linear function of predictors [19]. On the other hand, the actual values of the dependent variable are not a linear function most of the time. Although this algorithm seems very simplistic, it has a broad application in both conceptual and practical areas.

The most common questions which an analyst might seek for answers to them can be summarized in the following [20]:

- Is there a relationship between the response and predictor variables?
- How strong is the relationship?
- Is the relationship close enough to linear in order to deploy the linear regression algorithm?

- Which of the predictors contribute the most in predicting the response variable?
- How accurate the prediction can be for future cases?
- Is there any correlation between any of the predictor variables?

Equation (8.1) presents the simple linear regression model with a single predictor.

$$Y = \beta_0 + \beta_1 X + \epsilon \tag{8.1}$$

- Y: Response variable
- β_0: Constant value of the intercept
- β_1: Constant value of sthe lope
- ϵ: Error term following a $N(0, \sigma^2)$

Given the estimate of intercept $\left(\widehat{\beta_0}\right)$ and slope $\left(\widehat{\beta_1}\right)$, the values of the response variable can be predicted $(\widehat{y})$ as the Eq. 8.2 presents.

$$\widehat{y} = \widehat{\beta_0} + \widehat{\beta_1} x \tag{8.2}$$

- $\widehat{y}$: Predicted response variable
- $\widehat{\beta_0}$: Predicted value of the intercept
- $\widehat{\beta_1}$: Predicted value of the slope

Residuals or error term can be obtained using the Eqs. (8.3) and (8.4). Residual analysis has a wide application, especially for model validation techniques.

$$e_i = y_i - \widehat{y}_i \tag{8.3}$$

$$e_i = y_i - \left(\widehat{\beta_0} + \widehat{\beta_1} x\right) \tag{8.4}$$

The coefficients of a linear regression model can be obtained by minimizing the residual sum of squares (RSS) where there are n available observations using Eqs. (8.6) and (8.7).

$$\text{Min} \sum_{i=1}^{n} e_i^2 \tag{8.5}$$

$$\widehat{\beta_1} = \frac{\sum_{i=1}^{n} (x_i - \bar{x})(y_i - \bar{y})}{\sum_{i=1}^{n} (x_i - \bar{x})^2} \tag{8.6}$$

$$\widehat{\beta_0} = \bar{y} - \widehat{\beta_1} \bar{x} \tag{8.7}$$

Standard error (SE) of an estimator is very helpful in determining the confidence intervals for the estimators. Equations (8.8) and (8.9) can be used in order to calculate the standard error values, while σ^2 is the variance of the error term (ε).

$$\text{SE}\left(\widehat{\beta}_0\right)^2 = \sigma^2 \left[\frac{1}{n} + \frac{\bar{x}^2}{\sum_{i=1}^{n}(x_i - \bar{x})^2}\right] \qquad (8.8)$$

$$\text{SE}\left(\widehat{\beta}_1\right)^2 = \frac{\sigma^2}{\sum_{i=1}^{n}(x_i - \bar{x})^2} \qquad (8.9)$$

Confidence intervals for the estimators can be calculated based on Eq. (8.10) and cumulative normal table. $(1-\alpha)\%$ confidence interval can be defined as a range of values which the true value is included in that interval with $(1-\alpha)\%$ confidence.

$$\text{Confidence Interval for } \widehat{\beta}_i = \widehat{\beta}_i \pm Z_{(1-\alpha)}.\text{SE}\left(\widehat{\beta}_i\right) \qquad (8.10)$$

Standard errors (SEs) can also be used to conduct hypothesis tests on the coefficients. For instance, a hypothesis might test whether there is a relationship between the response and predictor variable or not. Suppose that the regression model includes a single predictor. Therefore, the following hypothesis can test to find out if there is any relationship between the response and predictor variables. The analyst can either reject or fail to reject the null hypothesis. If the null hypothesis is rejected, it can be concluded that there is a relationship between the response and predictor variables.

$$\begin{cases} H_0 : \widehat{\beta}_1 = 0 \\ H_1 : \widehat{\beta}_1 \neq 0 \end{cases}$$

t-statistic can be used to draw a conclusion regarding the abovementioned hypothesis. Statistical software usually offers the probability of observing any value greater than or equal to $|t|$. This value is called the P value [21]. For instance, in a 95% confidence level, any of the estimators should have the P value below $(1-0.95 = 0.05)$ in order to be statistically significant. Equation (8.11) shows a t-statistic with t-distribution and $(n-2)$ degree of freedom. Equation (8.12) is the formula to calculate the residual standard error (RSE) based on the residual sum of squares (RSS).

$$t = \frac{\widehat{\beta}_1 - 0}{\text{SE}\left(\widehat{\beta}_1\right)} \qquad (8.11)$$

$$\text{RSE} = \sqrt{\frac{1}{n-2} \text{RSS}} \qquad (8.12)$$

R-squared is one of the most commonly used statistics in order to make a judgment regarding how well a model performs. Indeed, it is a measure of how close the data are to the fitted regression line. It should be noted that the linear regression parameters are estimated by minimizing the distance between the data point and the regression fitted line. The higher value for the R-squared statistic is an indication of a model with more acceptable performance. R-squared is always between 0% and 100%.

$$R^2 = 1 - \frac{\sum_{i=1}^{n}(y_i - \widehat{y}_i)^2}{\sum_{i=1}^{n}(y_i - \bar{y})^2} \qquad (8.13)$$

R-squared adjusted is an adjusted version of the R-squared which takes into account the number of predictors used in developing the model. The R-squared value would not be decreased if more predictors are being added to the model. Therefore, R-squared adjusted has a broad application for the cases which the number of the predictors in each model is not the same [22]. Equation (8.14) is the R-squared adjusted formula where n and p are the sample size and number of the predictors, respectively.

$$R^2_{\text{adj}} = 1 - \frac{\frac{1}{n-p-1}\sum_{i=1}^{n}(y_i - \widehat{y}_i)^2}{\frac{1}{n-1}\sum_{i=1}^{n}(y_i - \bar{y})^2} \qquad (8.14)$$

The main difference between the linear regression model with single and multiple predictors is in the number of independent variables which may carry helpful information to predict the response variable [23]. Equation (8.15) shows the general regression equation with n independent variable. The rest of the terminologies and principles are as same as what has been mentioned before.

$$Y = \beta_0 + \beta_1 X_1 + \beta_2 X_2 + \ldots + \beta_n X_n + \epsilon \qquad (8.15)$$

8.2.1.1   Linear Regression Assumptions

1. Regression model should be linear in parameters. It means that the relationship between the response and predictor variables should be linearly explainable [24].
2. The mean residual should be zero. As mentioned earlier, the error term should have a mean value very close to zero. A hypothesis should be formed in order to accurately investigate this assumption.

3. Homoscedasticity of residuals or equal variance should be held. This assumption indicates that the error variance should not have any specific trend over the observation values. For instance, the variance of the error term should not increase as the observation value increases.
4. Residuals should follow a normal distribution.
5. There should not be any autocorrelation between residuals.
6. Predictor variables and residuals should not be correlated.
7. The number of observations should be greater than the number of predictors.
8. There should be variability within the predictor variables. It implies that a predictor should not always have close to the same value.
9. There should not be any multicollinearity between the predictors.

8.2.2 Logistic Regression

Logistic or logit regression is an adapted version of the linear regression which is able to perform the classification tasks when the dependent variable is binary. Indeed, logistic regression is a predictive model which studies the relationship between a binary variable and set of independent variables. The response variable can only have two classes as failure and success. Independent variables can have either binary or continuous values. Multinomial logistic regression is an extension of the logistic regression where the response variable can have more than two levels. Ordinal logistic regression is a multinomial logistic regression where the order of the classes is important [25].

Logistic regression models should be interpreted in a different way compared with the linear regression models. Logistic regression is the relatively mature and well-developed method. Logistic regression seeks to predict the probability of a class rather than predicting a number. It should be noted that the range of the linear regression prediction values is from minus infinity to infinity, while logistic regression prediction values are always between zero and one [26]. In order to apply the linear regression methodologies for logistic regression, odds should be predicted instead of probability. The measurement unit of the log of odds is called logit from the logistic unit. Odds is a ratio of the probability of an event happening over the probability of an event not happening. Odds can only have positive values from zero to infinity. Therefore, it is not still suitable for linear regression which offers minus infinity to infinity values. Natural Log function can be applied to be able to deploy linear regression principles [27].

$$\text{Odds} = \frac{P}{1-P} \tag{8.16}$$

$$\text{Ln(Odds)} = \text{Ln}\left(\frac{P}{1-P}\right) \tag{8.17}$$

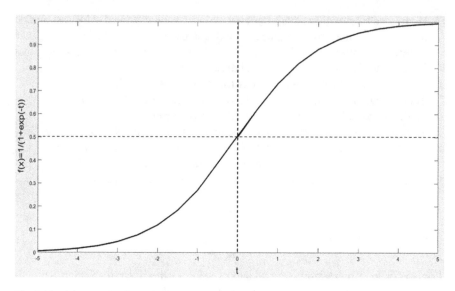

Fig. 8.10 An example of a logistic function

Equation (8.18) presents a linear regression model with two predictors, where the response variable is the log of the odds ratio. It should be considered that Eq. (8.20) is presenting a logistic function. In general, the logistic function is a sigmoid function which can take any real value. Figure 8.10 presents an example of a logistic function.

$$\mathrm{Ln}\left(\frac{P}{1-P}\right) = \beta_0 + \beta_1 X_1 + \beta_2 X_2 + \varepsilon \qquad (8.18)$$

$$\mathrm{Ln}\left(\frac{\widehat{P}}{1-\widehat{P}}\right) = \widehat{\beta}_0 + \widehat{\beta}_1 X_1 + \widehat{\beta}_2 X_2 \qquad (8.19)$$

$$\widehat{P} = \frac{e^{\widehat{\beta}_0 + \widehat{\beta}_1 X_1 + \widehat{\beta}_2 X_2}}{1 + e^{\widehat{\beta}_0 + \widehat{\beta}_1 X_1 + \widehat{\beta}_2 X_2}} = \frac{1}{1 + e^{-\left(\widehat{\beta}_0 + \widehat{\beta}_1 X_1 + \widehat{\beta}_2 X_2\right)}} \qquad (8.20)$$

As Fig. 8.11 depicts, binary classification can be viewed as a task of separating two classes by a linear separator in the feature space. In higher dimension space, the linear separator would be a hyperplane rather than a line. The model coefficients or weights should be determined based on the training data set. In order to test the model performance, a novel data set might be tested on the developed model [28]. If the developed model can be validated using the test and verification data sets, the future points can be classified based on the developed model. As mentioned earlier, the logistic regression is only able to handle the response variable with only two classes. The best linear model would be the one which has the least squared error. In practice, numerical methods such as gradient descent or conjugate gradient can be

Fig. 8.11 Schematic view
of logistic regression in
two-dimensional feature
space

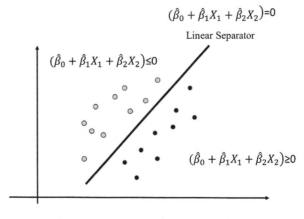

$$(\hat{\beta}_0 + \hat{\beta}_1 X_1 + \hat{\beta}_2 X_2) = 0$$

Linear Separator

$$(\hat{\beta}_0 + \hat{\beta}_1 X_1 + \hat{\beta}_2 X_2) \leq 0$$

$$(\hat{\beta}_0 + \hat{\beta}_1 X_1 + \hat{\beta}_2 X_2) \geq 0$$

Fig. 8.12 An example of
data points which are not
linearly separable

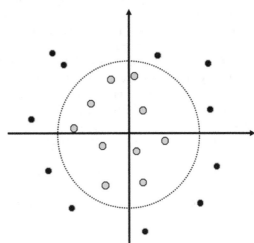

applied in order to learn the model coefficients. It has been highly recommended to use the normalized features to be able to directly compare the model coefficients.

The higher odds ratio of a feature indicates a stronger relationship of that feature with the target value. The odds ratio of one means not really related. The negative odds ratio is also indicating an inverse relationship between the feature and the target. Appropriate interpretation of the odds ratio is very critical. For continuous features, the odds ratio can be interpreted as changes in the ratio of the odds when the feature is increased by one unit [29].

It should be noted that it might be possible that there is not any linear separator for a data set, as Fig. 8.12 shows. It means that the data is not linearly separable. It looks like a nonlinear function might be able to separate the two classes. The challenge is how the logistic regression might be able to learn a nonlinear function. The overall idea is to use some mapping functions in order to use basic functions to map the features in higher dimensional space where the points are linearly separable [30]. For

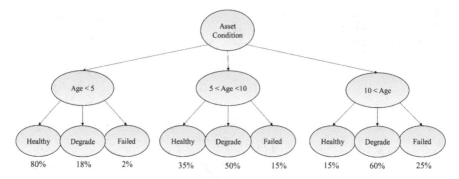

Fig. 8.13   An example of a decision tree

instance, in the example of Fig. 8.13, the data points are not linearly separable in terms of two features (x_1 and x_2). A basis function using quadratic terms ($x_1{}^2$ and $x_2{}^2$) might be able to solve this problem. Mapping functions are usually a polynomial function of the original features.

The main application of this model for asset management is in determining the probability of failure for an asset or system given the future condition. In addition to that, logistic regression might be very helpful for real-time advanced warning models which can quickly detect if any of the points fall in the high failure probability area. There are various types of ML and statistical algorithms which each has its own application. An analyst should cleverly select the most applicable model based on each problem. So far, linear and logistic regression have been discussed in this chapter. As a reminder, linear regression tries to develop a model which joins the data points with a minimum error, while logistic regression seeks to develop a model which separates the data points.

8.2.2.1   Logistic Regression Assumption

1. The linear relationship between the independent variables and a log of odds [31].
2. Observations should be independent than each other.
3. The dependent variable does not need to have a linear relationship with the independent variables. The reason is that the logistic regression applies nonlinear log transformation of the odds ratio [32].
4. The error term does not need to be normally distributed. But it still needs to be independent than other variables.
5. Homoscedasticity assumption is not needed. Logistic regression does not need variances to be heteroscedastic for each level of the independent variables [33].
6. There should not be high multicollinearity between the independent variables.
7. Logistic regression typically requires a large sample size.

8.2.3   Decision Tree

Decision tree uses a tree-like model of decisions using the possible consequences. The decision tree has wide applications in the fields of operations research (OR) and decision analysis. Decision trees are one of the most popular ML algorithms which seek to find the best strategy which most probably reaches to the ultimate goal [34]. In general, decision trees are similar to the flowcharts in terms of structure. Indeed, decision trees are a map of the possible outcomes of series of the choices which are all related. Each internal node represents a test on an attribute. Branches represent the outcome of the test, and leaf nodes stand the decision classes [35]. A decision tree always starts with a single node which branches to the possible ultimate outcomes. Decision trees try to find an algorithm to drive the optimum solution mathematically. The applications of the decision trees are for both regression and classification tasks. Each decision node indicates a possible outcome for the target. In order to draw a numerical analysis, the cost of each action and the expected outcome of each node should be defined. Decision trees should be continued until each path reaches an endpoint and no more choices remain to be made. An in-depth analysis is helpful to reach to the minimum risk and desired outcome. The decision-maker's preference should be taken into account when defining the desired outcomes. One of the main advantages of the decision trees is that the outputs of the algorithm are easily human interpretable [36].

An automated algorithm is needed if the training data is large. Algorithms which automatically build the decision trees are called decision tree induction or decision tree learning. These algorithms build the tree top-down by divide and conquer approach. One of the most important decisions in building the trees is the way in which the features should be selected [37]. A variety of heuristic algorithms such as ID3 system of Quinlan (1979) are available to gain more information at each level of building a decision tree. Given a training data set, several decision trees might be obtained based on the priority of the features at each level. Entropy approach has been suggested by Claude Shannon (1916–2001) who has been considered as the father of information theory. Equation (8.21) is presenting the entropy formula where p_0 and p_1 are the fractions of each binary class. It should be noted that regression trees allow continuous target at the leaves. Variance at the node leaf can be used instead of entropy.

$$\text{Entropy}(S) = -p_1 \, \text{Log}_2 \, (p_1) - p_0 \, \text{Log}_2 \, (p_0) \qquad (8.21)$$

Real-world data sets might have various features which increase the complexity of the trees. Indeed, what happens in the background is that there are various approaches in order to trim a tree to reduce its level of complexity. One of the most important criteria in building a decision tree is making a decision on when to stop a branch [38]. There are a variety of approaches in this regard. For instance, it should be always considered that the current value of the objective function would not improve in the next levels. Therefore, if a branch with the more desirable outcome already exists, that branch can be trimmed. Most of the time, the analyst set the maximum depth of a tree before building a decision tree. The maximum length refers to the length of the largest path. Another approach is pruning a tree by removing the features which have low importance for the model [39, 40].

8.2.4 Support Vector Machine

Support vector machine (SVM) is one of the common supervised algorithms which is able to perform the classification and regression tasks. SVM has been empirically shown to have acceptable performance on a variety of applications. In addition to the linear regression task, SVM is able to perform the classification for nonlinear problems using the kernel trick [41]. Kernel trick is able to map the input data to the higher dimensional space of the features. In other words, kernel trick uses complicated transformation techniques on labeled training data sets to separate the outputs. Therefore, given a labeled training data set, the SVM algorithm is able to define an optimal hyperplane which classifies the new examples. The main advantages of the SVM algorithm are its versatility for defining various kernel functions for each decision function as well as its effectiveness for high dimensional problems [42]. It should be considered that SVM might use very complicated computations and transformations which are not easily human interpretable. Figure 8.14 presents an example of 2D SVM with a linear separator. Instances close to the separator, shown by a circle, are called support vector points. The primary purpose of the SVM is maximizing the margin, ρ, which is the distance between the support vector points [43].

Distance from an example d to the separator can be calculated as follows:

Fig. 8.14 An example of 2D SVM with linear separator

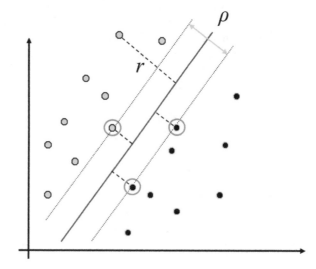

$$r = \frac{\beta_0 + \beta.d}{\|\beta\|} \tag{8.22}$$

$$\|\beta\| = \sqrt{\sum \beta_i d_i} \tag{8.23}$$

Two classes can be shown as Eqs. (8.24) and (8.25).

$$\beta_0 + \beta.d \geq 1 \qquad \text{for } t = +1 \tag{8.24}$$

$$\beta_0 + \beta.d \leq 1 \qquad \text{for } t = -1 \tag{8.25}$$

It should be noted that Eqs. (8.24) and (8.25) can be combined as Eq. (8.26).

$$t.(\beta_0 + \beta.d) \geq 1 \tag{8.26}$$

The following is the detail of the problem for a binary SVM classification. The objective is finding the weights, βs, which minimize the objective function.

Find β_0 and β such that

$$\min \ Z(\beta) = \|\beta\|^2$$

$$t_i.(\beta_0 + \beta.d_i) \geq 1 \quad \forall (d_i, t_i)$$

As Fig. 8.15 depicts, the classifier is only depending on the support vector points. It means that the same classifier would have been obtained if the support vector points were the only points of the training data set. This indicates that only support vector points are the matter for the classification task and the other points of the training data set might be ignorable. Indeed, that is the reason why SVM is resistant

Fig. 8.15 Sensitivity of the SVM algorithm to support vector points

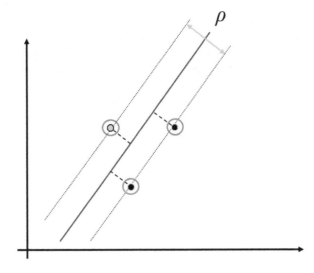

to overfitting. If the number of support vector points is getting close to the number of instances, overfitting might happen with a higher chance [44]. Figure 8.15 shows the sensitivity of the SVM algorithm to support vector points.

SVM and logistic regression are both seeking to find hyperplanes of the classifier by applying different search bias. Logistic regression provides a hyperplane which has the minimum error. On the other hand, SVM focuses on maximizing the margin rather than the error.

8.2.5   Naïve Bayes

Naïve Bayes algorithm is one of the commonly used families of the probabilistic classifier. Naïve Bayes algorithm is based on the Bayes' theorem which assumes strong independent relationships between the features. This independence is called naïve [45]. Naïve Bayes has wide applications in text recognition, spam detection, medical diagnosis, etc. The main competitor of the Naïve Bayes classifier can be more advanced methods such as support vector machine (SVM). In general, Naïve Bayes can be considered as a simple classifier technique which can assign the class labels, from a finite set to the problem instances [46]. The family of Naïve Bayes algorithms exists based on the same main principle. The core assumption of the Naïve Bayes algorithm is independent of a particular feature from any other features. For instance, an asset might be degraded if the efficiency, vibration, and temperature are not falling in an acceptable predefined range. Naïve Bayes classifier deliberates that each of these features independently affects the probability of the asset to be considered as degraded one without any possible correlation between the features [47]. In the various practical application of the Naïve Bayes algorithm, the parameters can be estimated based on the method of maximum likelihood. It means that the analyst might be able to apply the Naïve Bayes algorithm without accepting the Bayes probability theorem. It should be noted that the Naïve Bayes algorithm performs well on various complex real-world problems.

In general, the probability of an event, shown by *p(event)*, maps an event to a real number which is always between 0 and 1, with the probability of 0 and 1 for impossible and sure events, respectively. As mentioned earlier in this book, the random variable is a measurable value associated with an event. For instance, for throwing a fair coin, the random variable can have either tail or head. Conditional probability is one of the most important concepts of probability which can be considered as the basis of Bayes' theorem [48]. Conditional probability is an updated probability of an event given some other events have occurred. Equation 8.27 presents the general concept of the conditional probability of event A given event B.

$$p(A|B) = \frac{p(A \cap B)}{p(B)} \tag{8.27}$$

The main idea of the Bayes' theorem is to calculate the p(Hypothesis| Evidence) in terms of p(Evidence| Hypothesis) which is usually easier to estimate. Hypothesis could be asset health condition, healthy or faulty, and the evidences could be the features representing the condition of the asset. Equation 8.28 presents the Bayes' theorem which is based on the simple principles of conditional probability.

$$p(\text{Hypothesis}|\text{Evidence}) = \frac{p(\text{Evidence}|\text{Hypothesis}).p(\text{Hypothesis})}{p(\text{Evidence})} \tag{8.28}$$

Indeed, the analysts are trying to compute the p(Hypothesis| Evidence) using the simpler measures based on Bayes' theorem. From asset management point of view, p (Failure| Sympton) should be estimated using the right side of Eq. (8.28) which can be directly estimated from data. Two events are independent if happening of one event does not change the probability of the happening of the other one. Equations (8.29) are presenting the concept of the independency. Independency assumption is often used to simplify computations of complicated probabilities.

$$p(A|B) = \frac{p(A \cap B)}{p(B)} = \frac{p(A).p(B)}{p(B)} = p(B) \tag{8.29}$$

The joint probability distribution for a set of random variables $X_1, X_2, \ldots, X_n$ states the probability of every combination of values $p(X_1, X_2, \ldots, X_n)$. Probability of any event can be calculated given a joint probability distribution. Marginal distribution of a subset of a collection of random variables is the probability of variables contained in the subset. The probability of a subset of variables can be achieved based on the joint distribution function without referring to the values of the variable which are not contained in the subset [48]. Once the marginal distribution function is calculated, the conditional distribution can also be computed as Eq. (8.30) shows.

$$p(A|B, C) = \frac{p(A, B, C)}{p(B, C)} \tag{8.30}$$

In Naïve Bayes, the basic assumption is that the features are conditionally independent than the target value. Assume Y denotes for the response (hypothesis) variable, while X stands for a set of features (evidence). As Eq. (8.31) presents, this assumption states that the probability of a feature taking some values does not depend on values of other features given the target.

$$p(X_1|X_2, Y)) = p(X_1|Y) \tag{8.31}$$

The following equations are presenting the Naïve Bayes model derivation for a set of discrete features. Equations (8.33) and (8.34) have been obtained based on the definitions of marginal and conditional probability distributions.

$$p(Y|X_1, X_2, \ldots, X_n) = \frac{p(X_1, X_2, \ldots, X_n|Y) \cdot p(Y)}{p(X_1, X_2, \ldots, X_n)} \tag{8.32}$$

$$p(Y|X_1, X_2, \ldots, X_n) = \frac{p(X_1, X_2, \ldots, X_n|Y) \cdot p(Y)}{\sum_Y p(X_1, X_2, \ldots, X_n, Y)} \tag{8.33}$$

$$p(Y|X_1, X_2, \ldots, X_n) = \frac{p(X_1, X_2, \ldots, X_n|Y) \cdot p(Y)}{\sum_Y p(X_1, X_2, \ldots, X_n|Y) \cdot p(Y)} \tag{8.34}$$

Equation (8.35) has been developed based on the Naïve Bayes strong assumption.

$$p(X_1, X_2, \ldots, X_n|Y) \cong \prod_{i=1}^{n} p(X_i|Y) \tag{8.35}$$

Therefore, the Naïve Bayes models can be written as Eq. (8.36). Although Naïve Bayes does not generally work well for continuous targets (regression), it can be extended to continuous features using probability density functions instead of probability mass functions [49].

$$p(Y|X_1, X_2, \ldots, X_n) \cong \frac{\prod_{i=1}^{n} p(X_i|Y) \cdot p(Y)}{\sum_Y \prod_{i=1}^{n} p(X_i|Y) \cdot p(Y)} \tag{8.36}$$

8.2.6   Artificial Neural Network

The main concept of artificial neural network (ANN) has been inspired by the biological neural networks. ANN is not simply an algorithm which seeks to define a framework for ML algorithms to work together in order to process the complex data. Indeed, ANN seeks to work such as human or animal brain to learn from the past, collect experience, and improve future decisions [50, 51]. For instance, in the field of image recognition, thousands of training pictures might have been used to label an object to detect that specific object for future examples. ANN has gotten more attention since several layers have been applied to describe more complex systems. It seems that the ANNs are the most robust ML algorithm.

The base on an ANN is based on the connection between various units called neurons, which have been inspired by the biological neurons. Connection signals between artificial neurons are real numbers which are going to pass through either linear or nonlinear functions to generate the outputs. These connections between the neurons are called edge. The learning process is usually adjusted based on the

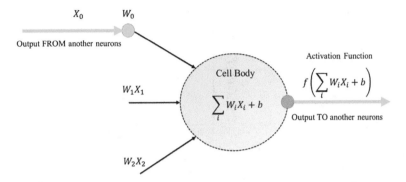
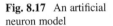

Fig. 8.16 An overall view of ANNs

Fig. 8.17 An artificial
neuron model

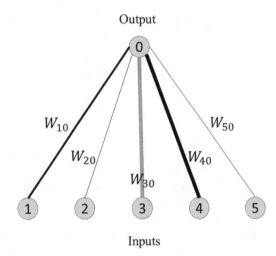

weights of the neurons and edges [52]. The human brain has roughly 100 billion neurons with an average of 10,000 connection for each of them. Figure 8.16 depicts an overall view of ANNs [53].

Figure 8.17 presents an artificial neuron model where the widths are representing weights. Black and red stand for positive and negative weights, respectively. Each neuron has a threshold to be activated or fired.

$$\text{Output} = \begin{cases} 0 & \sum_i W_i X_i < \text{Threshold} & (\text{Not} - \text{Activated}) \\ 1 & \sum_i W_i X_i \geq \text{Threshold} & (\text{Activated}) \end{cases}$$

A single artificial neuron is called perceptron. Perceptron uses iterative learning algorithms to obtain the best estimate of the weights in order to reach the optimal output given a training data set. Indeed, the weights would remain the same if the

Fig. 8.18 Weight values
versus training error

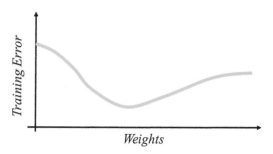

output is correct. The weights should increase (decrease) if the output is lower
(higher). Equation (8.37) presents an overview of perceptron learning rule, while
γ, t_j, O_j, and O_i are the learning rate, correct output value of neuron j, predicted
output value of neuron j, and predicted output value of neuron i, respectively. The
weight would be iteratively updated until the convergence is being reached. Learning rate is a value between 0 and 1 which too small or large values lead to very slow
learning and oscillation between inadequate solutions, respectively [54].

$$W_{ji(t)} = W_{ji(t-1)} + \gamma\left(t_j - O_j\right)O_i \qquad (8.37)$$

If the data points are linearly separable, the value set of the weights would
eventually cover a consistent value based on the perceptron theorem. If the data
points cannot be linearly separated, the cycling would happen which repeats a set of
weights through an infinite loop. The hypotheses are trying to find the best set of
weights which minimizes the training error [55]. Gradient descent is one of the well-
known algorithms which conceptually works as a hill climbing task. Indeed, the
values of the weights would be changed by a small amount to reduce the training
error. Figure 8.18 presents the weight values versus training error in
two-dimensional space.

ML algorithms based on the biological neurons can be categorized as two main
classes as the following present. It should be considered that the perceptron cannot
perform well if the data points are not linearly separable [56].

- Perceptron, which has been developed around the 1950s, is the initial algorithm
  for learning simple ANN. These simple ANNs usually have only a single layer.
- Backpropagation, which has been developed around the 1980s, is a more complex learning algorithm for multilayer ANN.

Multilayer networks connect multiple perceptrons to form an ANN with multiple
layers. A typical ANN consists of an input, hidden, and output layer which are fully
connected. Each node is similar to a perceptron, and function can be determined
based on the weights. Figure 8.19 presents a typical ANN with one hidden layer
[57]. Possible outputs are not only limited to one option. For instance, the possible
outputs could state that an asset condition is healthy, moderate, or critical. Nonlinear
activation functions are needed in order to move beyond the linear problems. It

Fig. 8.19 A typical ANN
with one hidden layer

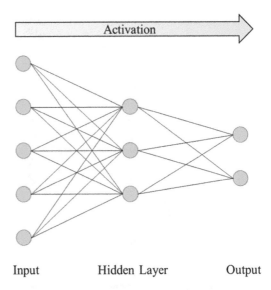

Input   Hidden Layer   Output

should be noted that a differentiable function is needed to apply the gradient descent approach. The logistic function is an example of a nonlinear output function. More complex systems probably need a greater number of hidden layers. ANN with many hidden layers is called "deep learning."

8.2.7 Deep Neural Network

As Fig. 8.20 depicts, ANN with many hidden layers are called deep. Deep learning or deep neural network (DNN) became one of the most popular learning algorithms especially in the most recently developed fields such as text recognition and face detection. One of the most obvious examples of the broad applications of DNNs is the ability of social media, such as Facebook, to automatically detect the faces in a picture and tag them based on the security restrictions [58]. Figure 8.21 presents an example of a face recognition task using a DNN with three hidden layers.

Each node one layer is connected to all the other nodes of the next layer in a fully connected neural network. This might end up to too many connections by considering the raw inputs, such as pixels of 1000 by 1000 image, as the main features. The first layer can have 1000 nodes, while the total number of connections can be as high as 1000,000,000 [59]. Therefore, it might be computationally intensive to reach an output. Pixel values as raw inputs might still have one more problem regarding the location of the objects in a picture. For instance, as Fig. 8.21 depicts, algorithms should be able to detect the face of a person or an object in a picture regardless of the location. Therefore, computational efforts and location dependency might be among

Deep Neural Network

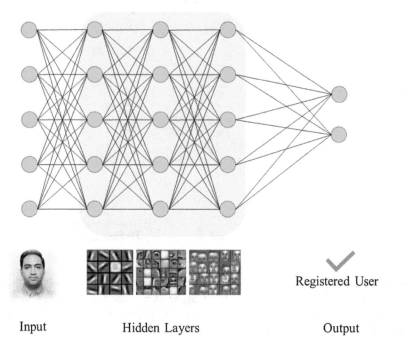

Registered User

Input   Hidden Layers   Output

Fig. 8.20 An example of a face recognition task using a deep neural network with three hidden layers

Fig. 8.21 Location independence feature of the DNN algorithms

the challenges of a DNN. Convolutional neural networks (CNNs) are able to overcome these problems [60].

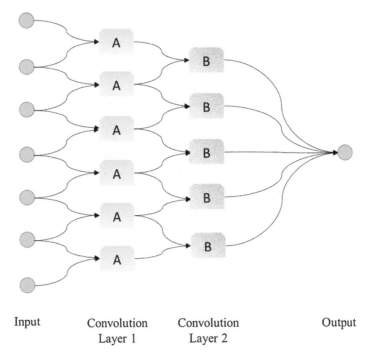

Input   Convolution   Convolution   Output
               Layer 1   Layer 2

Fig. 8.22 A typical CNN with two convolutional layers

8.2.8 *Convolution Neural Network*

Convolution neural networks (CNNs) are a class of deep neural networks. The main idea of the CNNs is to not fully connect the network and replicate the neurons at many places within a layer called convolutional layer [61]. Figure 8.22 presents a typical CNN with two convolutional layers. The same nodes or neurons "A" and "B" are representing all over the convolutional layers. The same neuron means that all the instances will have exactly the same weights in that layer. Since there are fewer connections due to the same neurons at multiple places, this neural network archi-tecture is less computationally intensive. Furthermore, the same neurons are used in multiple places and are able to generalize to other places as well [62].

CNNs are usually couples with pooling layers which can be viewed as subsampling or zooming-out feature. Indeed, the most beneficial information would pass to the next layer instead of all the information. The main benefit of the pooling layers is reducing computational efforts. Figure 8.23 represents a CNN with "pooling max" layer which only passes the maximum of the incoming nodes.

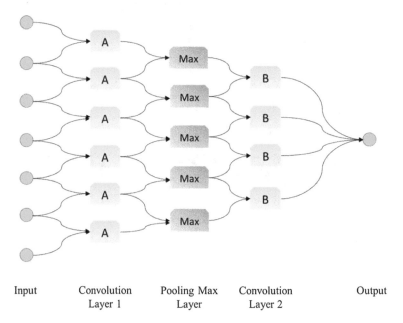

| Input | Convolution Layer 1 | Pooling Max Layer | Convolution Layer 2 | Output |

Fig. 8.23 A typical CNN with two convolutional layers and one pooling max layer

8.3 Machine Learning Algorithms Evaluation

The main goal of the ML algorithms evaluation techniques is to find out how well the algorithms perform when deployed. This indicates that the performance of the algorithms should be tested on a new set of data which has not been seen during the training phase. Evaluation techniques seek to estimate how the algorithms can be generalized by memorizing the training data in order to make a prediction for future points. A fundamental rule states that there should not be any overlap between the training and verification data sets since the performance of the model could be deceptively estimated well because it had already observed some of the points. Evaluation of the ML algorithm is highly depending on the application of the model. It should be noted that higher accuracy is not always desirable due to time and cost constraints. For instance, some healthcare applications such as cancer prediction algorithms cannot tolerate misestimation [63].

This should be considered that an unsuitable evaluation can lead to damage or misleading. In order to fairly assess the performance of the ML algorithm, suitable test example should be selected. Some algorithms might need to tune the value of few parameters for a given data set. Therefore, another part of the data, which is called validation or tuning set, should be separated to tune the parameter. ML algorithms are trained based on the training data set, and parameters of the algorithms are tuned by testing the performance of the algorithm on the validation data set. In the next step, the final developed model should be verified on the verification data set. There are some techniques such as K-fold to perform the repeated splitting task on data set. The following are a few examples of the most commonly used statistics to evaluate the performance of the ML algorithms for classification and regression tasks [64, 65].

- *Root mean square error (RMSE)* is usually used to compare the difference between the predicted values by models and actual or observed values. These deviations from the actual values are often called residual.

$$\text{RMSE} = \sqrt{\frac{\sum_{i=1}^{N} (\widehat{y}_i - y_i)^2}{N}} \tag{8.38}$$

- *Coefficient of variation (CV)* is a normalized value of the RMSE which facilitates the comparison of the models with different scales.

$$\text{CV} = \frac{\text{RMSE}}{\overline{y}} \tag{8.39}$$

- *R-squared* or *goodness of fit* measures how close the predicted data points are to the fitted regression line. *R*-squared is a value between 0% and 100% in which the higher the values, the better the model fits the data. It should be carefully considered that *R*-squared is not indicating whether a model is adequate or not. Indeed, residuals should be still analyzed beside the goodness of fit. Furthermore, higher values of *R*-squared do not necessarily indicate that the model performs well.
- *R-squared adjusted* is the modified version of the *R*-squared which considers the number of the parameters involved on a model.
- *Receiver operating characteristic (ROC)* is an example of statistics for classification task which is able to demonstrate the ability of the binary classification. The ROC curve plots the true-positive versus false-positive predictions.

8.4   Ensembles

So far in this chapter, various ML algorithms have been covered to develop a prediction model. Each of the predictive models could be developed based on a single ML algorithm. It might be possible to enhance the prediction task by building a combined model based on various ML algorithms. An ML model that combines several ML algorithms is called an ensemble or a committee which is supposed to overall work better than a single ML algorithm [66].

In ML and statistical analysis, ensembles can provide a predictive model which is a mixture of several models with higher performance than any of the constituent algorithms alone. Indeed, ensembles are combining several hypotheses to reach to a better one by applying the same base learner. Ensembles techniques can be categorized into supervised algorithms. It is expected that the performance of the ensembles be higher than the single algorithms since some algorithms might perform better on a part of data, while others perform better on the rest of the data. Random error caused by each algorithm can cancel out using the ensembles. In addition to that, each algorithm has a search and language bias which can be reduced by creating an ensemble including several algorithms. Performance evaluation of the ensembles usually needs more computational effort compared with the single algorithms. In order to speed up the calculation and convergence of the ensemble, a fast algorithm such as decision trees should be selected as the base learner. Ensembles might show more flexibility in predicting the response variable [67, 68]. Consequently, overfitting issue should be carefully analyzed before validating the results of the ensembles. Ensembles can be categorized as follows:

- *Homogeneous* ensembles apply the same algorithm while deploying the training data set to develop a model which is being established by various training data set.

  - *Bagging* ensembles take the repeated random samples.
  - *Boosting* ensembles assign a weight to the training data set. Weights are updated at each iteration to more focus on the latest correct predictions. Boosting techniques are mostly used for classification problems.

- *Heterogeneous* ensembles train multiple ML algorithms using the same data set.

8.5 Regression Example

The data for this example has been created by means of a numerical simulator of a naval vessel (frigate) characterized by a gas turbine propulsion plant [69, 70]. The main purpose of this example is to develop a model to predict the compressor and turbine degradation states, separately, by a combination of predictors. The following is a summary of the information for the regression example in this chapter.

- *Response variable*

  - CD: Compressor degradation estimates

- *Predictor variables* ($p = 16$)

  1. LP: Lever position (lp)
  2. SS: Ship speed (v) [knots]
  3. GTST: Gas turbine (GT) shaft torque (GTT) [kN m]
  4. GTRR: Gas turbine rate of revolution
  5. GGRR: Gas generator rate of revolutions (GGn) [rpm]
  6. SPT: Starboard propeller torque (Ts) [kN]

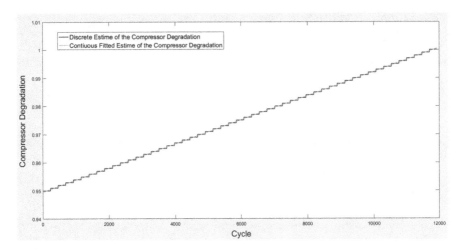

Fig. 8.24 Response variable of the regression example

  7. PPT: Port propeller torque (Tp) [kN]
  8. HPTT: High-pressure (HP) turbine exit temperature (T48) [C]
  9. GTCIT: GT compressor inlet air temperature (T1) [C]
 10. GTCOT: GT compressor outlet air temperature (T2) [C]
 11. HPTP: HP turbine exit pressure (P48) [bar]
 12. GTCIP: GT compressor inlet air pressure (P1) [bar]
 13. GTCOP: GT compressor outlet air pressure (P2) [bar]
 14. GTEGP: GT exhaust gas pressure (Pexh) [bar]
 15. TIC: Turbine injection control (TIC) [%]
 16. FF: Fuel flow (mf) [kg/s]

• *Number of observations (n = 11934)*

Figure 8.24 presents the response variable of the regression example. As mentioned earlier, the degradation index or healthy score might be a continuous or discrete variable. In this example, based on the physics of the compressor, the degradation mechanism follows a series of states. In order to meet the regularity assumptions of the multiple linear regression model, the response variable has been interpolated into a continuous variable. Interpolation is a method of creating new data points within the range of a discrete set of known data points.

Table 8.2 presents the results of the ANOVA table for regression example. Based on the *P* value criteria, the initial result indicates that the intercept, *GTCIP,* and *SPT* are not statistically significant for this example. It should be noted that the developed model in Table 8.2 only includes the predictors, which are statistically significant for the multiple linear regression model.

Various models have been applied to the given data set for the regression example. The presented results have been obtained based on the statistical and machine learning applications of the MATLAB software. The detail of the data set

Table 8.2 ANOVA table for linear regression model

```
Linear regression model:
    CD ~ FF + GTCIT + GTCOP + GTCOT + GTEGP + GTRR + GTST + HPTP + HPTT + LP + PPT + SS + TD + TIC
```

Estimated Coefficients:

	Estimate	SE	tStat	pValue
FF	0.58483	0.0091363	64.012	0
GTCIT	0.0082238	0.0005295	15.531	7.2005e-54
GTCOP	-0.056792	0.0011225	-50.592	0
GTCOT	-0.0028512	2.2258e-05	-128.1	0
GTEGP	1.1418	0.14756	7.738	1.0916e-14
GTRR	6.5151e-05	7.2343e-07	90.058	0
GTST	1.5558e-05	3.2625e-07	47.687	0
HPTP	0.40644	0.0067122	60.552	0
HPTT	-0.00085874	1.8634e-05	-46.083	0
LP	0.22042	0.0030928	71.269	0
PPT	-0.0030503	3.3794e-05	-90.263	0
SS	-0.057078	0.0010168	-56.134	0
TD	-1.2711	0.017071	-74.458	0
TIC	-0.00096623	1.3571e-05	-71.197	0

and generated MATLAB codes are provided in the appendix. Figures 8.25 and 8.26 present the true response (actual) versus predicted response value and response values versus residual, respectively, for various models. In this step, linear regression, stepwise linear regression, interaction linear regression, and decision tree are considered as the predictive model. Although more regularity assumptions are needed to be met to validate the results, in this step, it can be assumed that all the assumptions are met.

Table 8.3 presents the performance of the various ML algorithms based on the RMSE, R-squared, and training time in seconds. The linear regression model without the interaction terms does not perform satisfactorily compared with the other models. In terms of the RMSE and R-squared values, the performance of the stepwise and interaction linear models is more acceptable. It should be noted that interaction models are involving more predictors, which causes to increase the R-squared value. The adjusted R-squared statistic is more applicable to compare the performance of the models with the different number of predictors. The performance of the decision trees is roughly the same either with or without ensembles. The linear SVM model does not perform satisfactorily especially with high training time. Therefore, it can be concluded that the performance of the interaction linear regression is higher than other models for this given example.

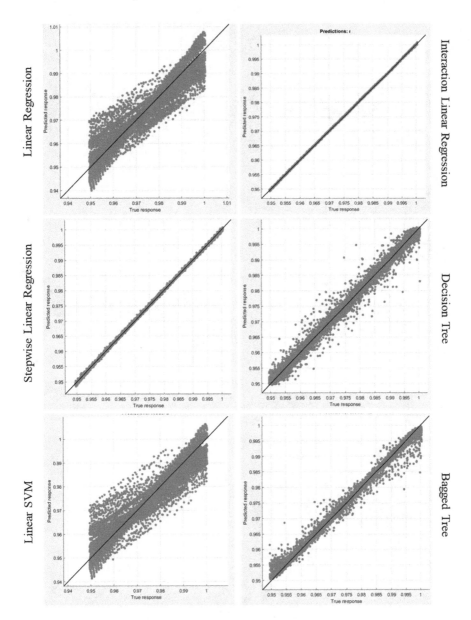

Fig. 8.25 True response (actual) versus predicted response value for various models

8.6   Classification Example

In this section, a classification example is presented. This example is investigating the stability of an electrical grid network. Stability of the electrical grid refers to the balance between the supply and demand. Traditional networks may achieve this

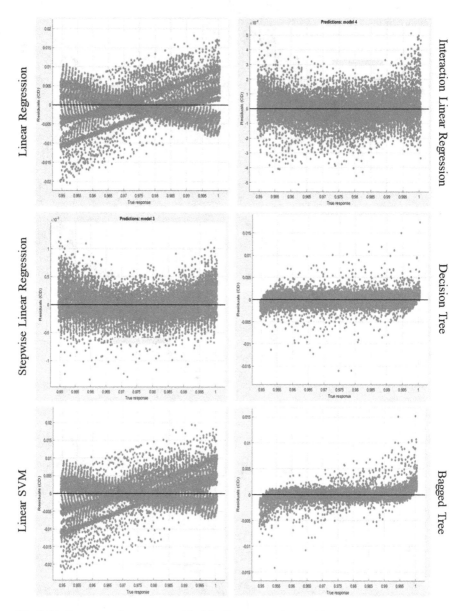

Fig. 8.26 Response values versus residual for linear regression model

balance by applying the demand-driven electricity production approaches. For future grids which the share of the renewable energy resources is higher, demand response strategies are promising. This indicates that electricity consumption may vary as the price of electricity changes. There are various ways to define the price of electricity through demand and supply forecasting methods. Decentral Smart Grid Control (DSGC) has received attention in order to investigate the electrical grid stability by

Table 8.3 Performance summary of the various ML algorithms

ML algorithm	RMSE	R-squared	Training time (sec)
Linear regression	0.0048	0.89	7
Stepwise linear regression	0.0002	1.00	889
Interaction linear	0.00009	1.00	9
Decision tree	0.0014	0.99	8
Bagged tree	0.0012	0.99	68
Linear SVM	0.0049	0.89	1253

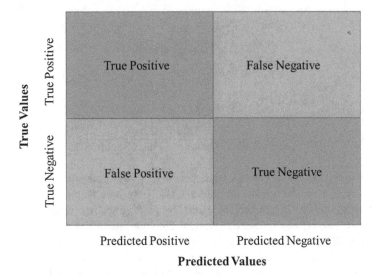

Fig. 8.27   An example of confusion matrix

considering more time-varying constraints. Various classification methods have been applied on the "Electrical Grid Stability Simulated Data Set," which has been donated to "UCI Center for Machine Learning and Intelligent Systems" by Vadim Arzamasov [71]. This example includes 10,000 instance, 12 predictors, and 1 binary response variable. The performances of the algorithms are compared based on the "accuracy" as a comparison statistic. Accuracy implies out of all predictions (e.g., positives or negatives) how many of them were correctly predicted.

The confusion matrix is one of the commonly used tools for classification problems. Confusion matrix has a specific table layout which visualizes the perfor-mance of an algorithm based on the true and predicted values of the response variable. Indeed, the confusion matrix seeks to find out if an algorithm may confuse the prediction task or not. Figure 8.27 depicts an example of a confusion matrix for a problem which the response variable has only two possible outcomes.

Equation (8.40) presents the definition of the accuracy statistic. Accuracy is used when neither class is more important than the other one. There are other statistics which might be used in order to compare the performance of the classification algorithms. Selection of the comparison statistics highly depends on the application of the study. Sensitivity and specificity are commonly used in medicine to report the prediction capability of a test. Precision and recall are typically used when the task is to extract some information. There is a trade-off between sensitivity and specificity as well as precision and recall.

$$\text{Accuracy} = \frac{\text{True Positive} + \text{True Negative}}{\text{True Positive} + \text{True Negative} + \text{False Positive} + \text{False Negative}}$$

(8.40)

$$\text{Sensitivity} = \frac{\text{True Positive}}{\text{True Positive} + \text{False Negative}}$$

(8.41)

$$\text{Precision} = \frac{\text{True Positive}}{\text{True Positive} + \text{False Positive}}$$

(8.42)

$$\text{Specificity} = \frac{\text{True Negative}}{\text{True Negative} + \text{False Positive}}$$

(8.43)

$$\text{Recall} = \frac{\text{True Positive}}{\text{True Positive} + \text{False Negative}}$$

(8.44)

When it comes to a classification problem, receiver operating characteristic (ROC) curve and area under the curve (AUC) are among the commonly used evaluating tools. The ROC curve is created by plotting the true positive against the false positive at various threshold settings. The true-positive rate is also known as sensitivity. The false-positive rate is also known as the probability of false alarm which can be calculated as (1 − specificity). ROC curve is a tool to select possibly the optimal models and to discard suboptimal ones independently (Fig. 8.28).

Area under the curve (AUC) is one way to summarize the performance of a predictive model in a single value which is commonly used for comparing predictive models. The higher AUC values present a better model, while the highest value of the AUC is 1. Figure 8.29 presents the application of the AUC through an example which seeks to compare the performance of predictive models A and B. Model which has the higher AUC value performs better than the other one.

ML algorithm	Accuracy (%)	Training time (sec)
Fine tree	84.2	10.9
Bagged tree	91.5	192
Fine KNN	81.1	23
Weighted KNN	86.6	39.6
Logistic regression	81.5	4.9
Linear SVM	81.5	48.3
Quadratic SVM	94.2	188.7
Cubic SVM	96.9	555.8

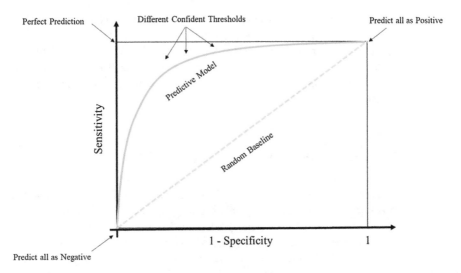

Fig. 8.28 An example of ROC curve

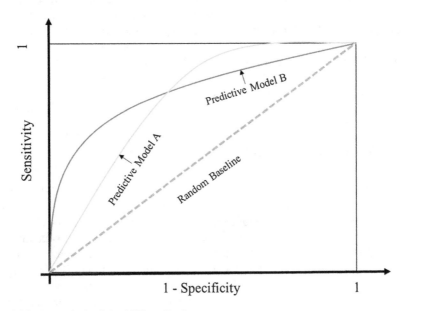

Fig. 8.29 An example of the AUC application

Cubic SVM has the highest accuracy, but the training time is also higher than other algorithms. Logistic regression is the fastest algorithm, but the accuracy is relatively low. It should be considered that the results of the ML examples, in terms of the algorithms performance, should not be overgeneralized to other cases. It means that each of these algorithms might perform differently in various applications (Figs. 8.30 and 8.31).

8.7   Machine Learning as a Service (MLaaS)

Machine learning as a service (MLaaS) includes various cloud-based platforms that cover most infrastructure matters such as data preprocessing, model training, retraining, and evaluation, with further prediction. MLaaS is a variety of services that provide machine learning tools as part of cloud computing services. MLaaS providers offer tools including data visualization, face recognition, natural language processing, predictive analytics, and deep learning. The providers' data centers are mainly handling the actual computation which might take place behind the user interfaces (UIs). The main benefit of these services is that customers can get started quickly with ML algorithms without having deep insight into the algorithms, install software or provision their own servers, just as same as other cloud services. "Azure Machine Learning by Microsoft," "Google's Cloud Prediction API," "AWS by Amazon," "Algorithms.io," "BigML," "Ersatz Labs," and "Nutonian/Eureqa" are a few examples of the cloud-based ML providers.

8.8   Concluding Remarks

Machine learning (ML) algorithms seek to extract the most beneficial information out of the raw data. In traditional algorithms, the analyst had to define the rules, which was not always accessible or easily possible, in order to obtain the output. ML algorithms develop the models or rules based on the training data set which include input and output data points. ML algorithms try to understand more beneficial information regarding the system based on the training data set. The built model can be tested using the verification data set which does not have any overlap with the training sets. If the model acquires the acceptable performance measures, it can be applied for other cases to perform the prediction process. Predictive models should be generalized which means performing satisfactorily for both training and verifying data sets.

The main purpose of this chapter was to present a few examples of the most commonly used statistical and ML algorithms more toward the application side rather than the theories behind the development of the algorithms. In this chapter, several commonly used supervised ML algorithms were presented in detail. A numerical example is also presented for regression and classification tasks in order

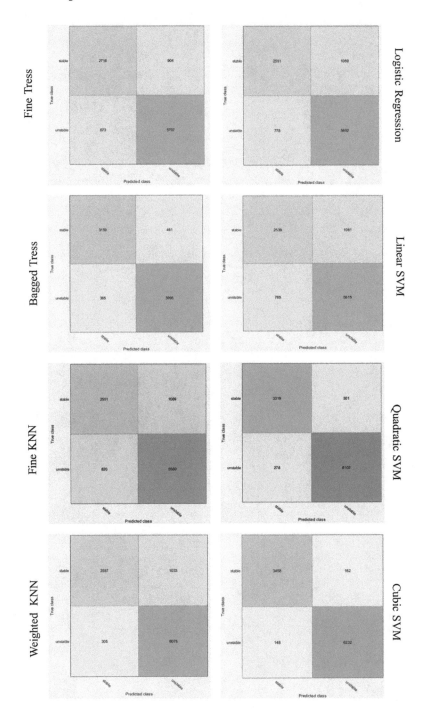

Fig. 8.30 Confusion matrixes for classification example

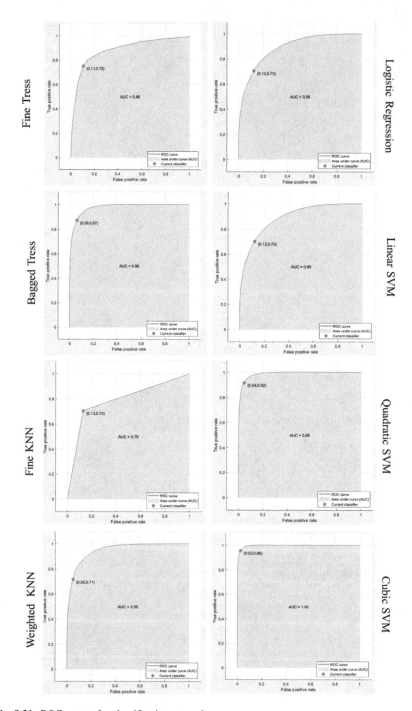

Fig. 8.31 ROC curves for classification example

to explain the performance of the various algorithms given the same data set. One of the most famous theorems in ML area is called "no free lunch." This theorem indicates that there is not an algorithm which can globally perform better than others for all the applications. Based on this theorem, several algorithm selection criteria were presented in this chapter.

References

1. D. Michie, D.J. Spiegelhalter, C.C. Taylor, *Machine Learning Neural and Statistical Classification*, vol 13 (1994)
2. I.H. Witten, E. Frank, M.A. Hall, C.J. Pal, *Data Mining: Practical Machine Learning Tools and Techniques* (Morgan Kaufmann, Amsterdam, 2016)
3. E. Alpaydin, *Introduction to Machine Learning* (MIT press, 2009)
4. K.P. Murphy, *Machine Learning: A Probabilistic Perspective* (MIT Press, Cambridge, MA, 2012)
5. S.B. Kotsiantis, I. Zaharakis, P. Pintelas, Supervised machine learning: A review of classification techniques, in *Emerging Artificial Intelligence Applications in Computer Engineering*, vol. 160, (2007), pp. 3–24
6. L. Bottou, Large-scale machine learning with stochastic gradient descent, pp. 177–186, 2010
7. P.M. Domingos, A few useful things to know about machine learning. Commun. ACM **55**(10), 78–87 (2012)
8. P. Baldi, S. Brunak, F. Bach, *Bioinformatics: the Machine Learning Approach* (MIT press, 2001)
9. M. Mohri, A. Rostamizadeh, A. Talwalkar, *Foundations of Machine Learning* (MIT Press, Cambridge, MA, 2018)
10. T. Mitchell, B. Buchanan, G. DeJong, T. Dietterich, P. Rosenbloom, A. Waibel, Machine learning. Annu. Rev. Comput. Sci **4**(1), 417–433 (1990)
11. E. Alpaydin, *Introduction to Machine Learning* (MIT press, 2014)
12. S. Marsland, *Machine Learning: An Algorithmic Perspective* (Chapman and Hall/CRC, 2014)
13. M.I. Jordan, T.M. Mitchell, Machine learning: Trends, perspectives, and prospects. Science **349** (6245), 255–260 (2015)
14. O. Chapelle, B. Scholkopf, A. Zien, Semi-supervised learning (chapelle, o. et al., eds.; 2006) [book reviews]. IEEE Trans. Neural Netw. **20**(3), 542 (2009)
15. T. Hofmann, Unsupervised learning by probabilistic latent semantic analysis. Mach. Learn **42** (1-2), 177–196 (2001)
16. R.S. Sutton, A.G. Barto, *Introduction to Reinforcement Learning*, vol 135 (1998)
17. R.S. Sutton, A.G. Barto, *Reinforcement Learning: An Introduction* (MIT Press, Cambridge, MA, 2018)
18. G.A. Seber, A.J. Lee, *Linear Regression Analysis*, vol 329 (Wiley, New York, 2012)
19. D.C. Montgomery, E.A. Peck, G.G. Vining, *Introduction to Linear Regression Analysis*, vol 821 (John Wiley & Sons, 2012)
20. S. Weisberg, *Applied Linear Regression*, vol 528 (John Wiley & Sons, 2005)
21. J. Neter, M.H. Kutner, C.J. Nachtsheim, W. Wasserman, *Applied Linear Statistical Models*, vol 4 (Irwin, Chicago, 1996)
22. M.L. King, Testing for autocorrelation in linear regression models: A survey, in *Specification Analysis in the Linear Model*, (Routledge, 2018), pp. 19–73
23. C.B. Santiago, J. Guo, M.S. Sigman, Predictive and mechanistic multivariate linear regression models for reaction development. Chem. Sci. **9**(9), 2398–2412 (2018)
24. A.F. Schmidt, C. Finan, Linear regression and the normality assumption. J. Clin. Epidemiol. **98**, 146–151 (2018)

25. D.W. Hosmer Jr., S. Lemeshow, R.X. Sturdivant, *Applied Logistic Regression*, vol 398 (John Wiley & Sons, 2013)
26. P.D. Allison, *Logistic Regression Using SAS: Theory and Application* (SAS Institute, 2012)
27. S. Menard, S.W. Menard, *Logistic Regression: From Introductory to Advanced Concepts and Applications* (SAGE, Los Angeles, 2010)
28. J.J. Arsanjani, M. Helbich, W. Kainz, A.D. Boloorani, Integration of logistic regression, Markov chain and cellular automata models to simulate urban expansion. Int. J. Appl. Earth Obs. Geoinf. **21**, 265–275 (2013)
29. J.C. Stoltzfus, Logistic regression: A brief primer. Acad. Emerg. Med. **18**(10), 1099–1104 (2011)
30. J. Starkweather, A.K. Moske, Multinomial logistic regression, Consulted page at September 10th: http://www.unt.edu/rss/class/Jon/Benchmarks/MLR_JDS_Aug2011.pdf, vol. 29, pp. 2825–2830
31. S. Sperandei, Understanding logistic regression analysis. Biochem. Med **24**(1), 12–18 (2014)
32. P.D. Allison, Measures of fit for logistic regression, 1–13
33. S. Menard, Standards for standardized logistic regression coefficients. Soc. Forces **89**(4), 1409–1428 (2011)
34. Y. Freund, L. Mason, The alternating decision tree learning algorithm, vol. 99, pp. 124–133
35. M.A. Friedl, C.E. Brodley, Decision tree classification of land cover from remotely sensed data. Remote Sens. Environ. **61**(3), 399–409 (1997)
36. T.K. Ho, Random decision forests, vol. 1, pp. 278–282
37. P.E. Utgoff, N.C. Berkman, J.A. Clouse, Decision tree induction based on efficient tree restructuring. Mach. Learn **29**(1), 5–44 (1997)
38. J.R. Quinlan, Induction of decision trees. Mach. Learning **1**(1), 81–106 (1986)
39. M. Mehta, J. Rissanen, R. Agrawal, MDL-based decision tree pruning, vol. 21, no. 2, pp. 216–221
40. R.E. Banfield, L.O. Hall, K.W. Bowyer, W.P. Kegelmeyer, A comparison of decision tree ensemble creation techniques. IEEE Trans. Pattern Anal. Mach. Intell. **29**(1), 173–180 (2007)
41. D. Meyer, F.T. Wien, Support vector machines, The Interface to libsvm in package e1071, pp. 28
42. T. Harris, Credit scoring using the clustered support vector machine. Expert Syst. Appl. **42**(2), 741–750 (2015)
43. B. Gu, V.S. Sheng, A robust regularization path algorithm for ν-support vector classification. IEEE Trans. Neural Netw. Learn. Syst **28**(5), 1241–1248 (2017)
44. S. Suthaharan, Support vector machine, pp. 207–235
45. T.R. Patil, S.S. Sherekar, Performance analysis of Naive Bayes and J48 classification algorithm for data classification. Int. J. Comput. Sci. Appl. **6**(2), 256–261 (2013)
46. I. Rish, An empirical study of the naive Bayes classifier, vol. 3, no. 22, pp. 41–46
47. A. McCallum, K. Nigam, A comparison of event models for naive bayes text classification, vol. 752, no. 1, pp. 41–48
48. G. Ridgeway, D. Madigan, T. Richardson, J. O'Kane, Interpretable boosted Naïve Bayes classification, pp. 101–104
49. H. Zhang, The optimality of naive Bayes. AA **1**(2), 3 (2004)
50. X. Yao, Evolving artificial neural networks. Proc. IEEE **87**(9), 1423–1447 (1999)
51. J.M. Zurada, *Introduction to Artificial Neural Systems*, vol 8 (West publishing company, St. Paul, 1992)
52. W.S. Sarle, Neural networks and statistical models (1994)
53. M.H. Hassoun, *Fundamentals of Artificial Neural Networks* (MIT press, 1995)
54. S.J. Russell, P. Norvig, *Artificial Intelligence: A Modern Approach* (2016)
55. M. van Gerven, S. Bohte, *Artificial Neural Networks as Models of Neural Information Processing* (2018)
56. P. Bangalore, L.B. Tjernberg, An artificial neural network approach for early fault detection of gearbox bearings. IEEE Trans. Smart Grid **6**(2), 980–987 (2015)

57. S. Shanmuganathan, Artificial neural network modelling: An introduction, in *Artificial Neural Network Modelling*, (Springer, Cham, 2016), pp. 1–14
58. G. Hinton, L. Deng, D. Yu, G. Dahl, A. Mohamed, N. Jaitly, A. Senior, V. Vanhoucke, P. Nguyen, B. Kingsbury, Deep neural networks for acoustic modeling in speech recognition. IEEE Signal Process. Mag. **29**, 82 (2012)
59. Z. Cai, Q. Fan, R.S. Feris, N. Vasconcelos, A unified multi-scale deep convolutional neural network for fast object detection, pp. 354–370
60. T. Do, A. Doan, N. Cheung, Learning to hash with binary deep neural network, pp. 219–234
61. M. Niepert, M. Ahmed, K. Kutzkov, Learning convolutional neural networks for graphs, pp. 2014–2023
62. O. Abdel-Hamid, A. Mohamed, H. Jiang, L. Deng, G. Penn, D. Yu, Convolutional neural networks for speech recognition. IEEE/ACM Trans. Audio Speech Lang. Process. **22**(10), 1533–1545 (2014)
63. N. Japkowicz, M. Shah, *Evaluating Learning Algorithms: A Classification Perspective* (Cambridge University Press, 2011)
64. S. Sra, S. Nowozin, S. J. Wright (eds.), *Optimization for Machine Learning* (MIT Press, 2012)
65. H.G. Schaathun, *Machine Learning in Image Steganalysis* (Wiley, Norway, 2012)
66. C. Zhang, Y. Ma (eds.), *Ensemble Machine Learning: Methods and Applications* (Springer, 2012)
67. R. Bekkerman, M. Bilenko, J. Langford (eds.), *Scaling Up Machine Learning: Parallel and Distributed Approaches* (Cambridge University Press, 2011)
68. G. Valentini, F. Masulli, Ensembles of learning machines, in *Italian Workshop on Neural Nets*, (Springer, Berlin, Heidelberg, 2002)
69. A. Coraddu, L. Oneto, A. Ghio, S. Savio, D. Anguita, M. Figari, Machine learning approaches for improving condition? based maintenance of naval propulsion plants. J. Eng. Mar. Environ **230**, 136 (2014)
70. Center for Machine Learning and Intelligent Systems UCI Machine Learning Repository, Condition based maintenance of naval propulsion plants data set
71. UCI Machine Learning Repository Center for Machine Learning and Intelligent Systems, Electrical grid stability simulated data data set, November

Chapter 9
Implementation Tools of IoT Systems

9.1 Introduction

IoT application is a complex software system that includes software components for devices, gateways, message aggregation, data analysis, storage, and visualization. As an emerging technology, IoT has been used in areas such as:

- Predictive maintenance [1, 2]
- Smart metering [3, 4]
- Asset tracking [5, 6]
- Connected vehicles [7–9]
- Fleet management [10]

These applications collect vast volume of telemetry messages from distributed IoT devices and processed them in real time in order to gain business insights and to take actions based on the insights. The sensitive data collected by IoT devices must be protected during data transmission and processing. Stable storage is needed to archive historical data. Scalable data analysis tools are needed to learn knowledge in real time from streaming data or as offline analysis from historical data. Data visualization and graphic user interface are needed to provide visual feedback and user control of data analysis and IoT devices [11, 12]. Each one of these components is a software program that requires significant engineering effort. Composing them so that they work together properly is even more challenging. In this context, developers of IoT applications may find it more productive to use existing IoT frameworks and other commercial off-the-shelf software components to build their final products.

© Springer Nature Switzerland AG 2020
F. Balali et al., *Data Intensive Industrial Asset Management*,
https://doi.org/10.1007/978-3-030-35930-0_9

9.1.1 IoT Platforms

In recent years, many IoT platforms have emerged to answer the demand for rapid prototyping of robust and scalable IoT applications. These IoT platforms include cloud-based systems such as AWS, Google, Azure, and IBM IoT [11–13] and stand-alone systems such as Node-RED [14], ThingWorx [11, 12], and ThingsBoard [15]. While differing in interfaces, implementation, performance, and costs, these IoT systems all offer device clients, messaging hubs, stable storage, data analysis modules, data routing mechanisms, and data visualization tools. From users' perspective, experience gained from working with one platform can be translated to others since these IoT platforms have similar design principles.

IoT applications are large software systems that are difficult to develop from the ground up. IoT platforms are designed to alleviate this burden. Despite the recent advances of IoT platforms, the complexity of IoT applications should not be underestimated. IoT applications have to be compatible with diverse and distributed sensors; able to process large volume of data in real time; tolerant to missing or erroneous data, runtime exceptions, and unreliable networks; scalable to big data analysis and storage; and extensible to complex workflow and business logic [11, 12]. As common to any complex software, utilization of appropriate software architecture and design patterns is essential to the success of IoT development projects. When making choices of the architecture and design patterns, it is important to understand the requirements of the IoT applications and balance them against the performance, limitations, and design trade-offs of the software components of an IoT platform [11].

9.1.2 Requirements of IoT Applications

Different types of IoT applications have different functional and nonfunctional requirements. Based on these requirements, developers need to select the suitable architecture, design pat- terns, and software components to be used on an IoT platform. Functional requirements refer to the specific functionalities of a software system, while nonfunctional requirements refer to the properties of the software application such as performance, security, fault tolerance, and cost.

For example, predictive maintenance and smart metering applications require device monitoring, abnormal event detection, and real-time and offline data analysis [1, 3, 4]. For applications such as asset tracking and fleet management, spatial locations of the sensor devices need to be maintained [5, 6, 10]. For connected vehicles, field gateways are often required to collect data from vehicles of each road segment and generate traffic alerts and send aggregate information to the cloud servers [7–9]. For connected vehicles, there are hard real-time constraints, which require low latencies [7–9]. For smart metering and asset tracking, high throughput is needed to process large volume of telemetry data [3–6]. For asset tracking and fleet management, offline analysis such as mixed integer linear programming is needed to find optimal scheduling [5, 10]. For predictive maintenance, offline analysis such as machine learning is needed to generate models that can be used to suggest proactive measures in real time [1]. All these applications require graphic user interface (GUI) for data visualization and control, user management, device management, device provisioning, and data security. While functional requirements often are related to individual IoT component, the nonfunctional requirements such as data security, latency, scalability, and fault tolerance are crosscutting concerns that affect the entire IoT system. Below is a summary of the functional and nonfunctional requirements of IoT applications.

1. Functional requirements

   (a) Real-time and/or offline data analysis.
   (b) Sensor data archive. Some types of data may be compressed.
   (c) Requirement and availability of gateways for IoT devices.
   (d) Real-time control of IoT devices.
   (e) GUI for data visualization and control.
   (f) Data processing workflow and/or business logic.
   (g) Distributed machine learning.
   (h) Device management and provisioning.
   (i) User management.

2. Nonfunctional requirements

   (a) Scalability to large number of IoT devices or telemetry messages
   (b) Strong data security
   (c) Hard real-time constraints with low latency or high throughput with soft real-time constraints
   (d) Tolerance of faults such as erroneous data and unreliable network
   (e) Cost requirement

9.1.3   Cloud-Based IoT Platform

IoT applications can be implemented using on-premise servers or cloud-based platforms. The choice mostly depends on the nonfunctional requirements of the application such as latency and security. Cloud-based software systems have gained tremendous growth in recent years with many legacy software systems migrated to cloud platforms to offer software as services rather than installation-based products. The benefits of cloud-based software systems include transparent updates, distributed and load-balanced processing, scalable computing and storage, and flexible offering of diverse products and services [16]. The drawback is that users of the cloud-based system need to maintain network connectivity for some product features to function correctly. Also, unstable network or latencies of slow network can cause unpredictable issues for applications with hard real-time constraints.

Fundamentally, a cloud platform is a network-connected distributed computing and storage servers that provide virtual environments for different operation systems, application containers (such as Google App Engine), and computing services [16]. A software system can run on a cloud platform by deploying its instances on virtual machines of the cloud platform though the advantage of this is limited to eliminating the need to acquire and maintain physical hardware. Users still have to set up the required computation environment such as dependency libraries, networking, and storage within the cloud servers. Application containers reduce some of these burdens in that the containers provide interfaces to computation environment that users do not need to separately set up for each deployment of the software system on a cloud platform.

There are four types of utilization of cloud-based platforms: IaaS, PaaS, FaaS, and SaaS with increasing level of dependency on the cloud platform [16–18].

1. IaaS (Infrastructure as a Service) is the ability of a cloud platform to act as virtual machines for software applications that normally runs in local servers. Amazon's EC2 (Elastic Computing Cloud) is such an example, where users can deploy software applications by launching instances of virtual servers, uploading the applications, and executing them. The virtual machines can be distributed and can use stable storage such as Amazon's S3 (Simple Storage Service) bucket.
2. PaaS (Platform as a Service) is a managed computation environment for specialized applications such as web servlets. Google *App Engine* is such an example, where users can develop software programs using *app engine* development tool and deploy the programs to the *app engine* to execute. In PaaS, users still need to develop complete software programs as in IaaS except that users rely on the API of the platform for runtime support, network, and storage.
3. FaaS (Function as a Service) such as *AWS Lambda* and *Azure Functions* allows users to implement lightweight applications as stateless functions, which can be used for high-throughput processing such as data transformation, filtering, and event detection. FaaS functions connect to storage and network through APIs. FaaS functions are usually considered serverless since there are no dedicated servers allocated for running the functions and the cost is based on calls to the functions.
4. SaaS (Software as a Service) refers to software applications hosted by cloud servers such as Dropbox that deliver functionalities over the network.

Most of the software components of a cloud-based IoT platform fit the mode of PaaS (e.g., Azure *event hub*, Cosmos DB, Redis Cache, and *Stream Analytics*), FaaS (e.g., *Azure func- tions*), or SaaS (e.g., Azure IoT Central). Azure IoT Central is a software service for developing IoT applications through a web interface, which is supposed to be simple to use with the least amount of control and customizability. PaaS and FaaS are used for customizable IoT applications that can be reasonably handled by the services such as *IoT hub* and *Stream Analytics*. Solutions based on IaaS are usually reserved to customized and complex IoT applications that require high level of control. Developers can utilize open-source software stack such as Spark, Mesos, Akka, Cassandra, and Kafka to build customized IoT applications and

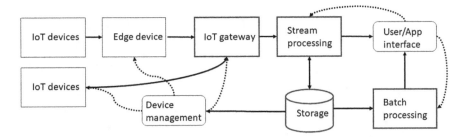

Fig. 9.1 Cloud IoT platform architecture, where solid line denotes direction of dataflow and dashed line denotes direction of control

deploy them in IaaS virtual machines. Since IaaS-based IoT applications need to reserve virtual machines, it is more suitable for computation-intensive applications with high rate of utilization of the virtual machines.

SaaS such as Azure IoT Central is intended for simple IoT applications with limited functionalities, while IaaS-based IoT applications are essentially locally developed applications deployed on IaaS virtual machines. Therefore, in this chapter, we will focus on cloud-based IoT components provided in forms of PaaS and FaaS.

9.2 Azure IoT Platform

Azure IoT platform refers to a suite of software tools offered on Azure cloud platform that can be used in combination to build IoT applications. While some software tools such as Azure IoT hub [19], *Stream Analytics* [20], and Power BI [21] are designed for IoT applications, others such as blob storage [22], Cosmos DB [23], and *Azure Functions* [24] are for general-purpose applications but frequently used in IoT application as part of the solution. The overall architecture of a cloud-based IoT system is shown in Fig. 9.1.

The architecture of cloud-based IoT platform is mostly driven by the functional requirements of IoT applications. The main requirements are to collect and analyze information collected from IoT sensors, to store the collected information, to report the results of analysis to users, and to integrate the analysis results with business systems [25]. The nonfunctional requirements such as security and latency are crosscutting concerns that impact multiple components of the IoT platform. In this section, we discuss each of the main components of Azure IoT platform, design choices, and related crosscutting concerns.

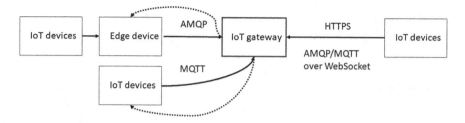

Fig. 9.2 Communication protocols between IoT devices and cloud gateway, where solid line denotes the direction of device-to-cloud messages and the dash line denotes the direction of cloud-to-device messages

9.2.1 IoT Devices and Gateways

The primary functionality of an IoT application is to enable IoT devices to send telemetry messages to a cloud gateway which enables further processing of the messages. To implement this functionality, a messaging protocol must be chosen for communication between IoT devices and cloud gateways. IoT client must be implemented for the IoT devices so that they can establish secure communication with a specific cloud gateway. In order to manage the IoT devices, an IoT application must also maintain storage for device identities, and device provisioning.

Messaging protocol IoT devices communicate with IoT gateways through messaging protocols such as MQTT, AMQP, and HTTP as shown in Fig. 9.2.

1. MQTT (Message Queue Telemetry Transport) is a lightweight client-server messaging protocol. It has a small memory footprint on the device and sends compact messages that use less network bandwidth [26]. Thus, MQTT is a popular choice for devices with limited computing resources and network connectivity. However, the compact messaging format also presents challenges to IoT applications. Since MQTT does not support message metadata in its header, IoT clients and gateways must agree on the format of the message content. MQTT has three levels of *quality of service* related to the message delivery: *at most once*, *at least once*, and *exactly once* delivery [27]. Azure *IoT hub* supports the first two but not the last one since it increases the latency while reducing the availability of distributed IoT infrastructure.

2. AMQP (Advanced Message Queuing Protocol) is a connection-oriented, multiplexing messaging protocol with compact message format, which is suitable for devices that require long-running connections and high-throughput communication with cloud gate ways [28]. Both AMQP and MQTT are implemented over TCP layer. While MQTT is more lightweight and suitable for devices that connect to cloud gateway directly with per-device credentials, AMQP allows more complex device topology so that multiple devices can connect with cloud gateway using the same secure connection.

3. IoT devices can also directly communicate with Azure *IoT hub* via HTTP though the implementation is more complex and the message overhead is higher than the binary format of MQTT and AMQP messages [28]. Moreover, cloud-to-device messaging is less efficient over HTTP since the device has to periodically poll messages, while AMQP and MQTT can push cloud messages to device with far less latency [28]. AMQP and MQTT can run over WebSocket for devices with restriction on open ports.

Device Client There are a variety of IoT devices with different processors, operating systems, networking capabilities, and runtime support of programming languages. For example, Intel NUC runs Ubuntu Linux, while Raspberry Pi can run Windows 10. For devices that run an operation system, IoT clients may be implemented using Azure IoT SDK [29] using various programming languages such as Java, C#, C++, Scala, Python, and JavaScript (Node.js). For the devices supported by Azure IoT SDK, the implementation of IoT clients is a relatively straightforward process that first establishes connection to cloud gateway with device credentials and then starts the loop of encoding messages and sending them asynchronously to the gateway. For FPGA-based devices such as CompactRIO, IoT clients can be developed using vendor-specific languages such as LabVIEW, though such client may not be compatible with specific cloud gateways such as Azure *IoT hub* due to its security requirement.

Cloud Gateway IoT devices may be connected to cloud gateway directly or connect through field gateways such as edge devices, which then connect to the cloud gateway. In the former case, IoT client program runs in IoT devices, while in the latter case, the IoT client runs in the field gateways. Field gateways may be used in cases where the IoT devices are low-power sensors that lack the computing power and network connectivity to reliably transmit IoT messages. Field gateways may be used to aggregate, filter, and preprocess raw telemetries from IoT sensors to reduce network congestion and workload of cloud gateway. Field gateway may also be used to connect to devices such as CompactRIO that lack suitable implementation of secure communication layers or to devices that can only be connected with industrial protocols such as OPC UA. In some cases, custom cloud gateway may also be developed using *Azure IoT protocol gateway* [30] to perform the functions of field gateway where the connection from IoT devices to the custom gateway may be secured using network tunneling such as VPN (virtual private network).

Azure platform supports two types of IoT gateways: Azure *IoT hub* and Azure *event hub* with available connection patterns shown in Fig. 9.3. Azure *IoT hub* is more suitable for applications that require bidirectional messaging between devices and cloud gateways [19]. Azure *event hub* is more suitable for applications that only need to send large number of messages to cloud gateway [31], and it does not support MQTT protocol. MQTT is based on publisher-subscriber model where message producers register as publishers in a MQTT broker and message consumers also register at the broker to listen on messages of selected topics. Azure *IoT hub* is such a broker except that MQTT protocol does not specify security requirement. To

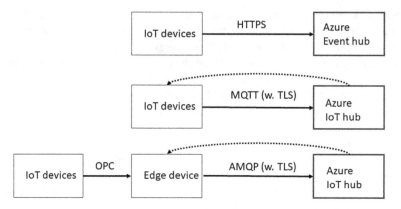

Fig. 9.3 Communication patterns between IoT devices and Azure IoT/event hub, where solid line denotes the direction of device-to-cloud messages and the dash line denotes the direction of cloud-to-device messages

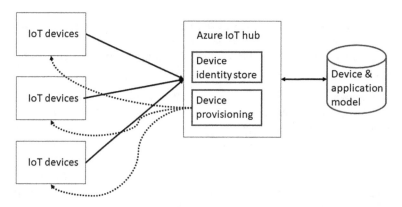

Fig. 9.4 Relation between IoT devices and device identity store, device provisioning service, and device and application model

protect the sensitive IoT data transmitted over untrusted network, Azure *IoT hub* requires connections be established over TLS (Transportation Layer Security) protocol [32]. In order to connect to *event hub*, the clients need to have the access keys associated with the *IoT hub* account that they will connect to. Applications can retrieve the messages through *IoT hub* using the same access keys.

After installing IoT clients on devices and setting up cloud gateway such as Azure *event hub*, developers need to make sure that only authorized devices can be connected, the messages from the devices can be understood by the rest of the application, and the devices can be managed and controlled remotely through cloud gateway and other components. To this end, an IoT application needs to have device identity store, device provisioning service, and device and application models as shown in Fig. 9.4.

While developers can configure small number of devices manually, for commercially produced IoT devices such as smart meters and smart thermostats, bulk processing of the device configuration and setup may be necessary. Azure *IoT hub* provides *device provisioning service* (DPS), which is a global service to support device registration and configuration [33]. DPS provides API for device configuration and registration using Azure *IoT hub*'s *device identity store* to provide per-device security credentials for authentication and authorization. DPS can also automate the registration of the devices to store the device model and application model.

The device model defines the schemas of device metadata, output, control parameters, and control actions. The application model contains the semantic relation between devices such as device topology and the relation between device and other application components. The device and application models are core design abstractions of each IoT application. For example, devices such as smart thermostats are modeled in relation to the room and building where they are located. A building management system may define a static hierarchical structure to model the devices based on their physical locations [6]. For applications such as fleet management, the device topology is more dynamic since the IoT devices may change their grouping as the vehicles carrying the devices are assigned to different tasks. Regardless of the specific designs, these models should be made persistent in storage such as SQL database or Azure Cosmos DB, which includes a graph API for convenient query of device topology.

9.2.2 Storage

IoT applications generate increasingly large amount of data that must be processed in real-time or by offline batch applications. Storing IoT data in on-premise database is not suitable for industrial IoT applications due to the lack of scalability of local storage. Cloud-based storage offers elastic capability that can scale up without significantly degrading performance [16]. Cloud-based storage can be divided into warm and cold storage with some trade-off between latency, query capabilities, and cost. Warm storage such as Cosmos DB and Azure SQL DB offers lower latency and flexible query interface but has higher cost. Cold storage such as blob storage and data lake has higher latency and simpler interface but lower cost. Figure 9.5 shows the possible dataflow path between cloud gateway, data processing computation, and storage. The distribution of data between warm and cold storage should be based on the requirement of IoT applications. To reduce cost, data that is not frequently used should be stored in cold storage. Data used in real-time processing components such as *Stream Analytics* and data visualization should be kept in warm storage [34]. A straightforward way of managing data for different storage is to store recently transmitted data in warm storage first and then gradually migrate old data into cold storage after a predetermined amount of time has passed.

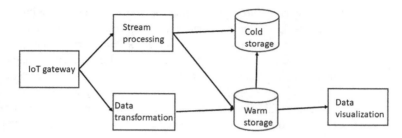

Fig. 9.5 Dataflow between IoT gateway, data transformation, stream processing, cold storage, warm storage, and data visualization modules

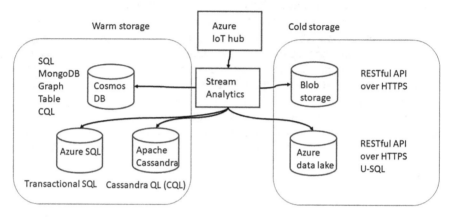

Fig. 9.6 Options of warm and cold storage with different query methods

There are a few options for warm and cold storage on Azure IoT platform as shown in Fig. 9.6. For warm storage, low latency is required. Azure provides Cosmos DB, which is a NoSQL database that supports five query methods including traditional SQL, MongoDB, graph, table, and Cassandra Query Language (CQL). Data in Cosmos DB can be set with an expiration date, after which the data is automatically deleted. Another option is Azure SQL DB [35], which provides relational database model with transactional support, which may be necessary for applications with strong integrity constraints. Both databases have flexible query interfaces, but they also have higher cost than that of cold storage such as blob storage and Azure data lake [36], which stores data as files or objects and uses RESTful API for query access. Query interface to cold storage is much more restrictive than that of warm storage. Thus, for computation such as visualizing power usage of a freezer over the past 30 days, warm storage should be used since the computation needs to issue queries that can return precise results with low latency. For computation such as analyzing the average power consumption of all freezers over the past year, cold storage can be used since the analysis can be computed offline with aggregate data set as input.

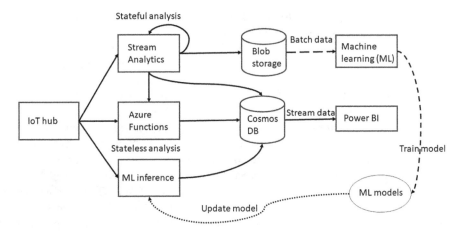

Fig. 9.7 Possible dataflow paths for IoT analytics

9.2.3   Data Analysis

The main purpose of IoT applications is to learn insight from data emitted from IoT devices and use the insight in making logical decisions toward business objectives [37] such as reducing operating cost of fleet vehicles or preventative maintenance of industrial equipment. Data analysis is the crucial step in realizing this goal. Depending on applications, IoT data analysis can be online processing of data streams or offline processing of data batches. For online processing, the computation can also be divided as stateless or stateful. Figure 9.7 shows the possible dataflow paths for the different types of IoT data processing on Azure platform.

For online processing of IoT data, latency is an important consideration [38]. Latency in this context refers to the amount of time it takes to analyze the IoT data collected over a time window (such as 5 seconds) to generate the input for a backend application such as a data visualization tool. For example, latency in the scale of seconds is not suitable for applications such as connected vehicles, which may have real-time constraints in milliseconds. Throughput is another consideration for online processing. Throughput refers to the amount of data that can be processed within a time period. Applications with high data ingestion rate require sufficient throughput to handle incoming telemetry data to avoid data loss or excessive latency. Insufficient throughput can increase latency since unprocessed data has to be buffered, and if the buffer overflows, unhandled telemetry data is dropped. An application can reduce the requirement of throughput by slowing down data ingestion rate using method such as sampling or aggregation. Cloud-based IoT platforms such as Azure can usually satisfy high-throughput requirements of IoT applications since user can allocate more resources to process data in parallel. However, latencies of these platforms are more difficult to reduce. For applications that demand near real-time response time, on-premise data processing is more appropriate. For example, field gateways can be used to perform preliminary data integration, data analysis, and aggregation before sending the lower-frequency data to the cloud gateway [39] such as Azure *IoT hub*. This also can reduce the computation cost associated with cloud platform usage and the network bandwidth.

Online data processing includes data transformation, data filtering, event detection, alert generation, data aggregation, and other streaming data analysis such as Key Performance Indicator (KPI) calculation, classification, and anomaly detection. Online data processing may be stateless or stateful computation. Stateless computation such as data transformation only needs its input data at time t to obtain its output at time t. Stateful computation needs to use both its input at time t and its states at previous time stamps such as $t - 1$ to compute its output and states at time t.

Azure includes two types of data processing systems: *Azure Functions* and Azure *Stream Analytics*. Both systems can take streaming inputs from IoT/event hubs and output results to Azure storage and reporting tool such as Power BI. *Stream Analytics* implements query computation on data streams, while *Azure Functions* provides general-purpose computation triggered by events. A common design pattern is to use *stream analytics* to preprocess data streams from *IoT hubs* and then forward the results to *Azure Functions* for advanced processing.

Azure *Stream Analytics* provides a domain-specific way for processing IoT data streams using a SQL-like (Stream Analytics) query language [40] that treats data streams as tables, stream events as records, and stream event fields as record fields. For example, select *V, I* from Power where *V* > 10 returns a stream of values that are the fields (*V, I*) from the stream Power where value of V is greater than 10. Nested queries can be made using *with* clause to define result sets of inner queries to be used in the outer query. In this query language, streams can be joined with the addition of an *on* operator that specifies the time bound of the joined stream events. The time bound is calculated with the *datediff* function using the time stamps of the stream events. Selected stream events can be *group*ed *by* field names with the addition of a *window* function or system time stamp. A *window* function specifies the grouping of stream events based on the event time stamps so that aggregate function can be applied to the events in each time window to output a single event. Azure *Window* functions include *tumbling window* (fixed-sized and nonoverlapping), *hopping window* (with fixed overlap), *sliding window* (all distinct windows with fixed size), and *session window* (events grouped by similar time stamps) [41]. This query language also provides a list of built-in functions to support computation on scalars, aggregates, records, spatial values, and temporal values (e.g., lagged events, last event, and first event in an interval).

Azure *Stream Analytics* can invoke user-defined function (UDF) [42] written in JavaScript in its queries. UDF allows users to perform operations such as math, array, regexp, and remote data access, which is not supported by built-in functions of the query language. *Stream Analytics* can also utilize user-defined aggregate (UDA) [43] that specifies how values are (de-)accumulated over event sequences grouped by window functions in a query. UDA is defined as a function that constructs a JavaScript object with fixed-named methods. *Stream Analytics* can also use *Azure Functions* to perform additional processing by sending query results via HTTP request to an Azure Function in batches [44]. The batch size is limited to 256 KB, while each batch is limited to 100 events by default. Note that Azure *Stream Analytics* can perform both stateless and stateful data processing. Through *window* and *analytic* functions, *Stream Analytics* can perform stateful computation on event streams subject to time-based constraints.

Azure *Stream Analytics* jobs can be launched through Azure's web interface by filling in the query statements in a textbox and specifying the input and output sinks of the analytics jobs. Launching a *Stream Analytics* job takes a few minutes, while error messages are few and uninformative, which makes it difficult to perform testing since mistakes in query statement such as typo of field names may simply yield empty results. Azure *Stream Analytics* does, however, allow faster tests by executing the queries on sampled data sets though this kind of testing does not reveal all possible faults with the queries. Also, the UDF and UDA associated with each *Stream Analytics* query are in separate scopes, which prevents sharing across multiple queries.

Azure Functions provide *serverless* computation triggered by events. *Azure Functions* can be implemented in languages such as Java, C#, and JavaScript to provide stateless computation in response to events from sources such as HTTP, blob storage, event hub, Cosmos DB, storage queue, and service bus [24]. *Azure Functions* allows the use of general-purpose programming languages to define complex logic to process event data and redirect the result to a destination such as blob storage. For stateful data processing, users can use durable *Azure Functions* [45], which is an extension of *Azure Functions* that implicitly uses Azure storage to manage stateful computation. As shown in Fig. 9.8, durable *Azure Functions* includes *orchestration* function that can compose multiple *activity* functions to run in sequence or in parallel. Durable *Azure Functions* also includes *orchestration* client that can start a long-running process that can be monitored for progress via HTTP endpoints.

As an evaluation of the throughput and latency of high-volume event processing with *Azure Functions*, Microsoft conducted a performance test [46] to demonstrate some of the optimization strategies. This test partitioned a stream of weather and seismic event data into two streams to be processed independently. The stream processing used Azure *event hub* to act as buffers and *Azure Functions* to separate

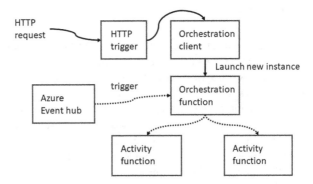

Fig. 9.8 Stateful data analysis using durable *Azure Functions*

the two types of event data and to push them into their respective event hubs for further processing. In this test, a stream of 100,000 events per second was processed for 9 days. During the test, half of the events were processed with about 1 second of latency, while 90% of the events were processed with less than 3 seconds of latency. However, max latency was about 4 minutes. The test achieved this level of throughput by configuring each *event hub* with 100 partitions, which translates to 100 virtual machines that execute the *Azure Functions* in parallel. This test also processed the streaming events in batches instead of individual events in *Azure Functions*, which increases throughput by reducing the overhead of launching *Azure Functions*.

Offline data processing includes KPI calculation using historical data, clustering, regression, and classification using unsupervised or supervised machine learning. Many of these offline data analysis tasks can be completed in Azure HDInsight [47], which is a PaaS cluster platform for launching distributed computing services. Apache Spark, which is a distributed in-memory programming library, is available in HDInsight as a computation environment for big data analysis [48]. With Spark, users can develop parallel applications that can perform data analysis using data from storage facilities such as Azure data lake and blob storage and gateways such as *event hub* (through Spark streaming). Spark utilizes a MapReduce programming model to support distributed computation, and its memory-based resilient distributed data set (RDD) abstraction provides better performance on big data than disk-based distributed file system as Hadoop does [49]. Spark supports offline analysis such as machine learning. For example, using Spark, an IoT application, can train a regression model offline using historical data and use the model to make online predictions for streaming data as shown in Fig. 9.7.

Note that Spark uses a cluster of machines to complete its tasks. When running Spark in HDInsight, a cluster of VMs of a minimum configuration such as a master and a slave must be created, and the cluster will exist until the task completes. Keeping a cluster running is far more expensive than services such as *Azure Functions*. Thus, it is not cost-effective to keep a long-running data analysis using Spark in HDInsight, which is more suitable for offline data analysis that is both computation-intensive and has a finite duration.

9.2.4   User Interface

Some IoT application such as smart thermostat can use the results of data analysis to generate automatic control signals such as setting the thermostat temperature. However, for most applications, graphic user interface is needed to render data visualization and generate reports for business decision-making. The interface can also be used to monitor the status of IoT devices and domain abstractions such as buildings that contain smart meters, to schedule an offline data analysis such as machine learning on historical data of certain time period, and to change an online data analysis such as the formula of a KPI. Azure IoT platform provides several PaaS and SaaS components for this purpose.

Visualization Azure Power BI is a versatile dashboard tool for data visualization and report generation of static or low-frequency data. Using the graphic interface of Power BI, non-programmers can create dashboard with various data charts from a range of data sources. Power BI includes two versions: Power BI Desktop and Power BI service (online). Power BI Desktop has more functionalities in terms of visualization and report generation. However, only Power BI service supports real-time streaming, and its data ingest rate is limited to five requests per second for its streaming data set and one request per second for push data set [50]. Higher-frequency data may be streamed or pushed to Power BI in batches with some added latency. For visualization and analysis on IoT data as time series, Azure *time-series insight* [51] may be used. *Time-series insight* is also a type of warm storage that stores data in memory and solid-state drives though its data is only kept for a limited amount of time and its query latency is still too high for on-demand applications.

Web and Mobile Apps Azure *app service* provides environment for building web app, mobile backend, and RESTful APIs. Users can create a web interface using Azure *app service* to connect to IoT components such as Azure *IoT hub* through WebSocket. For mobile devices with unreliable network and limited energy budget, push notification may be used to send IoT data such as alerts and event notification. To this end, Azure provides *notification hub* [52], which allows applications to push messages to mobile devices of platforms such as iOS and Android. Azure *notification hub* can also be integrated with Azure *app service mobile apps*, which works by retrieving device PNS (platform notification system) handles, registering the devices with notification hubs, and sending notifications from app backend through notification hubs.

9.3 Experiences with Azure

The Azure cloud offers a wide range of services, where each service belongs to a broader category to help with organization. As of this writing, there are 19 categories, where the number of services each category has ranges from as low as 2 to as high as 39. Many services are unique to each category, but some services do overlap and are placed in other categories as well. Information and examples of each service can be found at Microsoft's main Azure website. The wide range of services, categories, and amount of information online creates a steep learning curve to those unfamiliar with the Azure platform. To study the platform and ease of use, a set of tasks were selected to be performed on a data set. To illustrate the implementation of an IoT application using Azure platform, an experiment was conducted using Azure to analyze a set of time-series data of Li-ion batteries. The following sections detail the data set, the process, and our experiences using Azure. The experiences are based on the current iteration of Azure, which is always changing. Some features or services may not be available at the time this is read.

9.3.1 Data Set

The data set used details the operational use of a number of Li-ion batteries collected over an extended period of time by NASA [53]. This experiment used the data of battery number 10. The data was collected over many cycles, each of which includes a full charging phase followed by a full discharging phase, while metrics such as current, voltage, temperature, sampling time, and the impedance were collected. As the number of cycles increased, the time to fully discharge the battery decreased. The data was originally in MATLAB data files and was converted to CSV (comma-separated value) files before it is used for the simulation of IoT telemetries in the experiments.

9.3.2 Azure Setup

The main goal of this experiment is to create an IoT system using Azure. The experiment simulated telemetry data from sensors connected to the battery and sent it to an Azure *IoT hub*. The data is stored, processed, and then analyzed during each cycle. The analysis will then reveal when the battery should be replaced as its capacity degrades over time. The list of different services used during the setup is as follows: *IoT hub*, storage accounts, Stream Analytics jobs, Function App, Machine Learning Studio, resource groups, and cost management + billing (Fig. 9.9).

To use any Azure service, a user must have an account along with a subscription. The subscription keeps track of what services were used and how much to charge. A

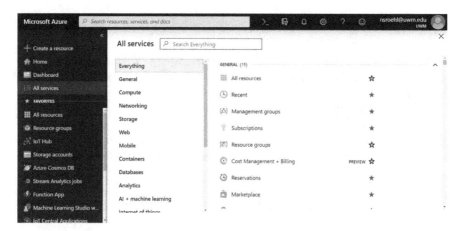

Fig. 9.9 View of the Azure portal. The left most menu contains favorite services. When clicking on any of these tabs, a window opens to the right. These windows are called blades. The *All resources* tab was clicked, showing only a small portion of all available services

nice feature is that many of the services in Azure only charge a user when they are running. If a user is looking to cut costs, turning off a service during low-peak hours is a viable option. The user still must be careful as some services may incur charges just for creating a resource.

9.3.3 Resource Groups

Before going into detail about many of these services and what was done, the importance of resources and resource groups must be stressed. When using any of these services, a resource from that service must first be created. For example, when using the *IoT hub*, a user must create an *IoT hub* resource first (Fig. 9.10). Since multiple resources can be created from the same service, a unique name is given to each one to distinguish them. Once a resource is created, it must be assigned to a resource group. Resource group is a convenient way to store multiple resources together in one place (Fig. 9.11). When creating some resources, Azure may create other resources unaware to the user. The resources created automatically will appear in the same resource group, keeping related resources together. This makes it easier to locate, delete, and manage multiple resources. Creating a resource can be done via command line inputs using the *Cloud Shell* or through the *Azure portal*. Both work through a browser and allow a user to create and monitor resources, but the *Cloud Shell* requires the user to know all the commands where the *Azure portal* is an interactive graphic interface. The use of *Azure portal* is described in this chapter.

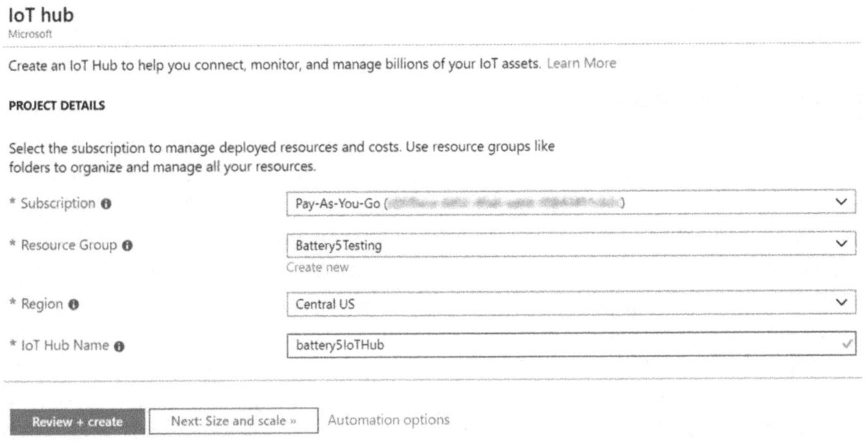

IoT hub
Microsoft

Create an IoT Hub to help you connect, monitor, and manage billions of your IoT assets. Learn More

PROJECT DETAILS

Select the subscription to manage deployed resources and costs. Use resource groups like folders to organize and manage all your resources.

* Subscription ❶	Pay-As-You-Go ( )
* Resource Group ❶	Battery5Testing
	Create new
* Region ❶	Central US
* IoT Hub Name ❶	battery5IoTHub

Review + create | Next: Size and scale » | Automation options

Fig. 9.10 Creating an *IoT hub*

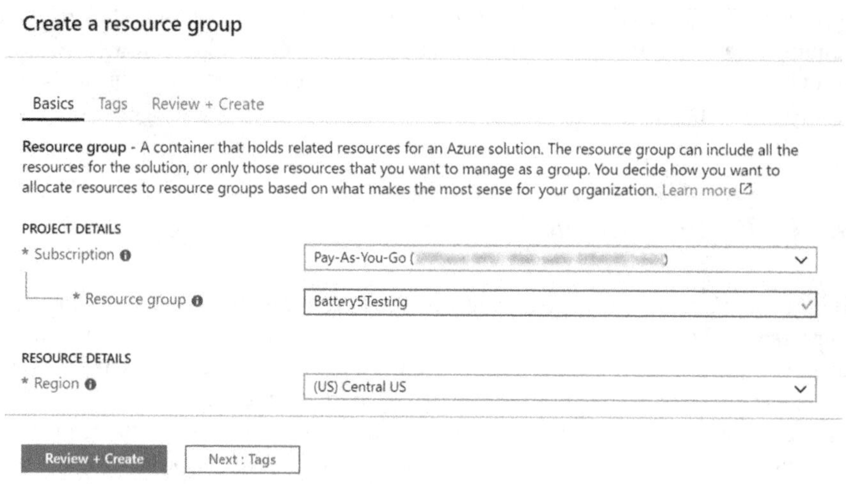

Create a resource group

Basics   Tags   Review + Create

Resource group - A container that holds related resources for an Azure solution. The resource group can include all the resources for the solution, or only those resources that you want to manage as a group. You decide how you want to allocate resources to resource groups based on what makes the most sense for your organization. Learn more ⌗

PROJECT DETAILS

* Subscription ❶   Pay-As-You-Go (⬛⬛⬛⬛⬛⬛⬛⬛⬛)   ⌄

  └── * Resource group ❶   Battery5Testing   ✓

RESOURCE DETAILS

* Region ❶   (US) Central US   ⌄

[Review + Create]   [Next : Tags]

Fig. 9.11 Clicking the resource groups tab on the left opens up the following blade to create a resource group. The process is very straightforward

Tier Name	Messages per day per unit	Number of units allowed	Cloud-to-device Messaging	IoT Edge Enabled	Device Management	Cost (USD) per month per unit
F1: Free Tier	8000	1	Y	Y	Y	$0
B1 : Basic Tier	400,000	200	N	N	N	$10.00
B2: Basic Tier	6,000,000	200	N	N	N	$50.00
B3: Basic Tier	300,000,000	10	N	N	N	$500.00
S1: Standard Tier	400,000	200	Y	Y	Y	$25.00
S2: Standard Tier	6,000,000	200	Y	Y	Y	$250.00
S3: Standard Tier	300,000,000	10	Y	Y	Y	$2,500.00

Fig. 9.12 Table of different tiers for *IoT hubs*. IoT Edge enabled means this tier can send/receive messages from IoT Edge devices. Device Management means this tier can use device twin, query, direct method and jobs [54]

9.3.4   IoT Hub and Storage

After creating a resource group, an *IoT hub* resource is needed as a location to receive messages [19]. An *IoT hub* can also send messages, but this functionality is not required for this experiment. When creating an *IoT hub* resource, a tier must be selected to specify how many messages this hub is expected to receive. Tier selection helps Microsoft allocate physical hardware resources on their end to meet the user's needs. The free tier was used in this experiment, as it is sufficient for this experiment. See Fig. 9.12 for more details on tiers. IoT devices must be added to the *IoT hub* after an *IoT hub* resource is created. Each device requires its own device ID in the *IoT hub* to distinguish one device from another. To add a device, select the *IoT Devices* tab under the newly created *IoT hub* resource. Here a user can add devices where Azure will auto-generate connection strings for each device. The connection string is critical in maintaining a secure connection between the device and the *IoT hub*.

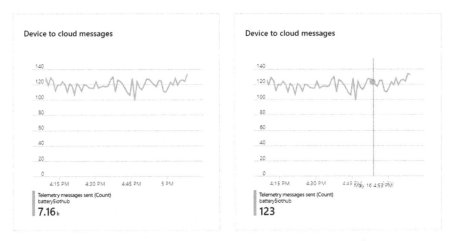

Fig. 9.13 Real-time graphs showing incoming messages to the *IoT hub*

To send messages from a device, Microsoft provides many SDKs and example code to connect various types of IoT devices to the *IoT hub* [55]. Various programming languages and platforms such as Java, C, .NET, Node.js, Python, and iOS frameworks are all supported. This experiment used Java, which needed Microsoft SDKs and Maven. It is recommended to use Maven as a build tool for Java-based IoT client, as Microsoft provides a dependency snippet that can be copied and pasted into the Maven build file (commonly named as pom.xml) which will then import all the appropriate library dependencies (namely, jar files) to connect to the *IoT hub*. Modifying one of the example codes provided by Microsoft was an easy way to start a project and send messages. The default connection string given in the example must be updated to match the devices ID in the *IoT hub*; if it is not, the messages will not send. To simulate telemetry messages, the battery data was first read from the CSV file, converted to strings in JSON format, and sent as MQTT messages to the *IoT hub* using the libraries provided. A quick way to view whether the messages are being sent or not is to examine the real-time graphs that are shown on the overview page of the *IoT hub* (Fig. 9.13). There will be some latency, but if the messages are sent correctly, the graphs will eventually update.

The account kind option determines what services can be used to store data. StorageV2 allows storage in blobs, files, queues, and tables. StorageV1 is similar to StorageV2 except that it is used for legacy accounts. Blob storage is the third option, but Azure documentation recommends not using it and opting for StorageV2 instead [56]. The replication option determines where duplicated data is stored geographically. Azure creates multiple copies of all data, the further away from the source, the more expensive storage gets, and replicated data is updated synchronously.

First storage option is locally redundant storage (LRS). LRS maintains three replicas in the same storage scale unit, which is a collection of storage racks. LRS is the most cost-effective way, but it provides the least durability in case of hardware failure or natural disasters. Zone-redundant storage (ZRS) maintains three replicas in the same data center but in different zones. If only the storage scale unit that contains

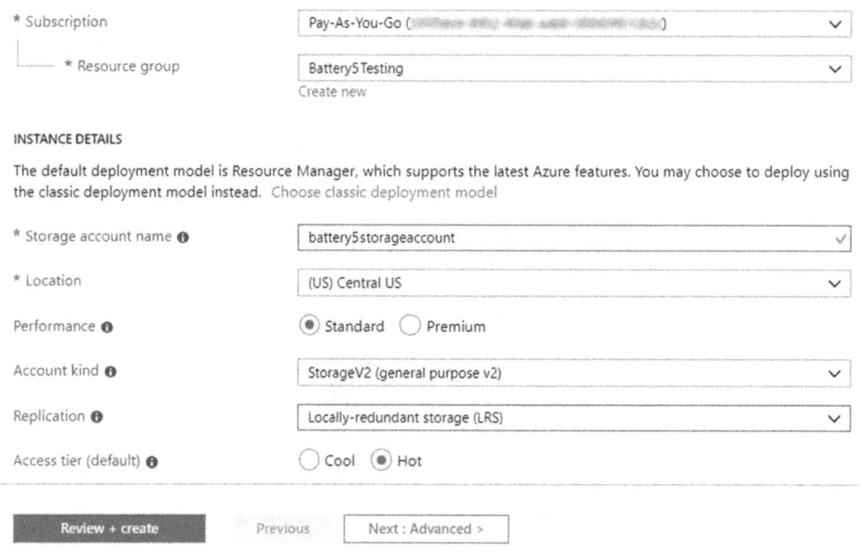

* Subscription	Pay-As-You-Go ( )	⌄

      * Resource group

Battery5Testing	⌄

Create new

INSTANCE DETAILS

The default deployment model is Resource Manager, which supports the latest Azure features. You may choose to deploy using the classic deployment model instead. Choose classic deployment model

* Storage account name ❶	battery5storageaccount	✓

* Location	(US) Central US	⌄

Performance ❶   ⦿ Standard   ◯ Premium

Account kind ❶	StorageV2 (general purpose v2)	⌄

Replication ❶	Locally-redundant storage (LRS)	⌄

Access tier (default) ❶   ◯ Cool ⦿ Hot

[**Review + create**]   [Previous]   [Next : Advanced >]

Fig. 9.14 Creating a storage account

source data malfunctions, replicated data will still be accessible with ZRS. This is not the case with LRS as the source and replicas are in the same storage scale unit. The last two options are geo-redundant storage (GRS) and read-access geo-redundant storage (RA-GRS). Both store three replicas in the same data center and three replicas at a different data center in a region hundreds of miles away. RA-GRS is the only option that allows read-only access to replicated data. Latency could be a problem with RA-GRS, as replicas at different data centers are updated asynchronously unlike updates done in the same data center. Lastly, cold or hot storage can be selected (Fig. 9.14). Hot storage should be used for frequent reads and updates. Use cold storage otherwise.

After creating a storage account with the appropriate settings, messages sent to the *IoT hub* can now be stored. If no other services are needed and storage is the only concern, the *IoT hub*'s Message Routing options can be updated to output data into four locations. They are event hubs, service bus queue, service bus topic, and blob storage [57]. Having a message reroute to an event hub seems redundant, since the *IoT hub* is essentially an event hub with bidirectional communication, but if a user wanted to combine information from an *IoT hub* and an existing *event hub* that collects data from multiple non-IoT sources for data analysis, this could be a useful option. A service bus is a message broker, where the *IoT hub* would be the publisher and any number of Azure services could be a subscriber [58]. A service bus queue stores messages in a single queue where a subscriber service reads from the queue when ready. Typically, service bus queues are meant for point-to-point communication. A service bus topic is similar to a service bus queue except that multiple queues are used. This allows multiple subscriber services to read from one source and is more similar to the typical publisher/subscriber model. If the data isn't needed immediately, blob storage can be used for long-term storage or for analysis at a later time.

When using the *Azure portal*, certain storage options don't allow a user to view the contents of stored messages. It is recommended to download and install Microsoft Azure Storage Explorer. This program is free and links up with an existing storage account after providing the correct credentials. After installation a user can now access all stored messages in their storage account from their desktop without having to go through the *Azure portal*.

9.3.5 Stream Analytics

If more destinations for storage or processing are needed, *Stream Analytics* is another service for routing messages sent to the *IoT hub* [59]. The basic setup of a *Stream Analytics* job are the inputs, query, and outputs (Fig. 9.15). Inputs are differentiated between data streams and reference data [60]. Data streams are unbounded streams of data, like *event hubs*, *Blob storage*, or an *IoT hub*. Reference data is typically static data that is used for correlation and lookups. Currently only *SQL databases* and *blob storage* are offered as reference data inputs. The query language used by *Stream Analytics* is similar to SQL and considered a subset of Transact-SQL (also known as T-SQL) [40]. Queries can determine or limit what gets stored and can be combined with windowing to perform analytics over designated periods of time. Output is where the data gets stored after being processed by the user-defined query [61]. Options for output are *event hub*, *SQL database*, *blob storage*, *table storage*, *service bus topic*, *service bus queue*, *Cosmos DB*, *Power BI*, and *Data Lake Storage Gen1*. *Azure Functions* is also an option for output, but these details will be discussed later. Unlike the *IoT hub*, *Stream Analytics* jobs must

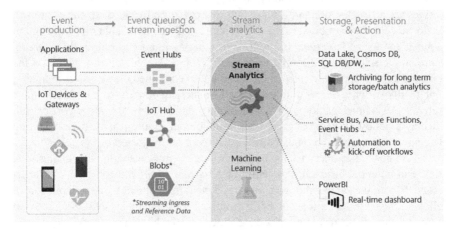

Fig. 9.15 Overview of inputs and outputs for *Stream Analytics* [59]

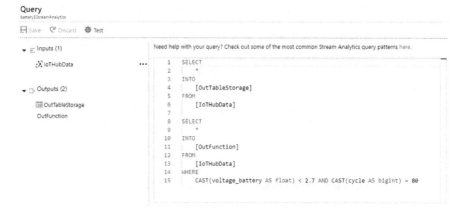

Fig. 9.16 *Stream Analytics* query with multiple parts. First selection takes all data sent to an *IoT hub* aliased as IoTHubData and outputs to *table storage* aliased as OutTableStorage. The second selection reads data from the same *IoT hub*, except when the parameters of the where statement are true, it executes an *Azure Function*

be started to be used. A start button is visible after creating a *Stream Analytics* resource; click this to start a job after all inputs, queries, and outputs are added. It may take a few moments for the job to start, but it will inform the user when it has started. A user is only charged when a job is running, so remember to stop the job when it is not in use.

When adding inputs or outputs, an alias is used in a *Stream Analytics* query to refer to an actual resource. Multiple inputs and outputs can be used for a single *Stream Analytics* job, reducing the amount of services required. For the experiment, a *Stream Analytics* job was created that used a *IoT hub* resource as input and a *table storage* resource as output. See Fig. 9.16 for example. Several other storage options could have been selected, but *table storage* provides all the functionality required. Tables are a structured NoSQL data store that can store lots of data, can scale with the data, and can be queried easily and quickly, and they are viable input and output sources for *Azure Functions* [62, 63]. When adding a table as output for a *Stream Analytics* job, it will ask for a Partition and Row key as field values which form the primary key for each row entry in the table. When combined they must be unique; if not it will overwrite the existing data with the matching primary key. Both keys must reference a named datatype in the streaming data. If not *Stream Analytics* will display a yellow warning symbol next to the output saying it is degraded. During testing no messages were updated when a *Stream Analytics* job became degraded.

Azure does offer query testing and some syntax error detection which helps save time and potential headaches as starting up a *Stream Analytics* job can be cumbersome [64]. In the same window, a query is edited; right clicking on the input source allows a test file to be loaded. Files to upload must be in JSON, CSV, or Avro format. Running the test after uploading, if done correctly, will give results to what is expected to happen during execution. See Fig. 9.17 for sample output. The testing

Results

outtablestorage   outfunction

Generated the Following:

- outtablestorage with 63 rows.

Download results

MESSAGEID	BATT_NAME	CYCLE	DATETIME	AMB_TEMP	VOLTAGE_B...	CURRENT_...	TEMP_BAT...	VOLTAGE_L...	CURRENT_L...	TIME
0	"B0005"	1	"4/2/08 1...	24	4.191491...	-0.00490...	24.33003...	0	-0.0006	0
1	"B0005"	1	"4/2/08 1...	24	4.190749...	-0.00147...	24.32599...	4.206	-0.0006	16.781
2	"B0005"	1	"4/2/08 1...	24	3.974870...	-2.01252...	24.38908...	3.062	-1.9982	35.703
3	"B0005"	1	"4/2/08 1...	24	3.951716...	-2.01397...	24.54475...	3.03	-1.9982	53.781

Results

outtablestorage   outfunction

Generated the Following:

- outfunction with 1 rows.

Download results

MESSAGEID	BATT_NAME	CYCLE	DATETIME	AMB_TEMP	VOLTAGE_B...	CURRENT_B...	TEMP_BATT...	VOLTAGE_L...	CURRENT_L...	TIME
58	"B0005"	80	"4/2/08 1...	24	2.652326...	-2.01071...	30.8597146	2.735	-1.9982	1056.922

Fig. 9.17 Results of testing query in Fig. 9.16

that is offered is helpful, but it is not foolproof. Queries can still be syntactically correct but produce no output. Azure does not provide any debugging tools to understand why something is not working. It is also unclear whether testing works with windowing, as windowing reads messages based on the time they were received and whether they reside inside a designated time slot (Fig. 9.18).

9.3.6 Azure Functions

Once the data is stored, an *Azure Function* can be triggered to read and perform some computations on the data. The main parts of an *Azure Function* are the trigger, bindings, and code implementation. The trigger causes a function to execute, and a function can only have one trigger [62]. Previously it was mentioned that *Azure Functions* can be used as output for a *Stream Analytics* job. The output is not a data stream but an HTTPTrigger that is sent when a certain condition is met to execute the function. Bindings can be distinguished by input vs. output. They are not required, but they provide an easier way to connect to other data sources for input or output in the function. See Fig. 9.19 for a list of some of the triggers and bindings offered.

Azure Functions can be written in C#, JavaScript, Java, and Python [65]. All languages listed can be written in an integrated development environment (IDE) on a desktop and then uploaded to Azure. An IDE is provided in *Azure portal* through the

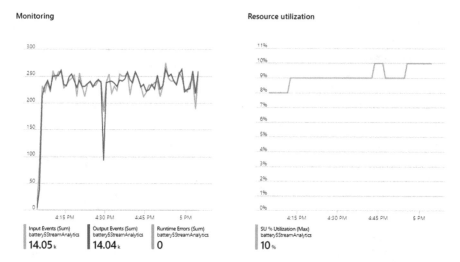

Monitoring   Resource utilization

Fig. 9.18 Real-time graphs showing messages being processed by *Stream Analytics*

Type	Trigger	Input	Output
Blob Storage	X	X	X
CosmosDB	X	X	X
Table Storage		X	X
Queue Storage	X		X
HTTP & Webhooks	X		X
Service Bus	X		X
Event Hubs	X		X
Timer	X		

Fig. 9.19 *Azure Functions* trigger, input, and output options [62]. Other services are available but were not included as they do not relate to IoT

browser, but it only supports C# scripts and JavaScript. Each option has its advantages and disadvantages. Using a desktop IDE gives the developer the choice to use a setup and programming language they are more familiar with. A desktop IDE also offers debugging and syntax error detection. Developing in the portal offers only two programming languages and no debugging or syntax error detection. Mistakes are only found when the code is executed. A downside to using a desktop IDE is having to set up the programming environment to be able to connect to Azure. This means having to go through Microsoft's online documentation which can be difficult to follow. Desktop IDEs also require a user to upload their functions to Azure, where anytime a function is updated, it must be re-uploaded. The portal avoids both of these issues.

To create a function in the portal, use the *Function App* service to create a *Function App* resource. The resource creation stage is where the programming language is selected. All functions that reside in this resource will use that language.

Fig. 9.20 The trigger and input/output bindings of an Azure Function

function.json

```
 1  {
 2    "bindings": [
 3      {
 4        "authLevel": "function",
 5        "name": "req",
 6        "type": "httpTrigger",
 7        "direction": "in",
 8        "methods": [
 9          "get",
10          "post"
11        ]
12      },
13      {
14        "type": "table",
15        "name": "inputTable",
16        "tableName": "battery5TableStorage",
17        "take": 50,
18        "connection": "AzureWebJobsStorage",
19        "direction": "in"
20      },
21      {
22        "type": "table",
23        "name": "outputTable",
24        "tableName": "battery5CoulombTable",
25        "connection": "AzureWebJobsStorage",
26        "direction": "out"
27      }
28    ],
29    "disabled": true
30  }
```

For the experiment, the .NET framework was chosen to create a C# script because it did not require much code, many of the better examples online were in C#, and ease of environment setup. The rest of the setup described is based on the user selecting the .NET framework in the portal to develop a function. The other programming language options were not tested and will not be discussed in great detail. Once the resource is created and opened in a new blade, a function can be made by clicking the plus symbol. After selecting the trigger to be used and providing a name, Azure will generate a C# script template and a function.json file. For script files, a function.json file is required to list the trigger and all input and output bindings [62, 66] (Fig. 9.20).

The processes of updating the JSON file and connecting to sources outside the function are the most difficult aspects when creating an Azure Function. The idea was very simple for the experiment: have a function read data from a table, perform some calculations on the data, and then output the new information to another table. One would think having the storage and processing all done in Azure would be a simple process, but sadly it is not. Microsoft provides a lot of documentation and examples online for how to add input and output bindings, but the amount of information and lack of explanations for each example given make it more difficult than it needs to be.

To update the bindings, select the *integrate* option underneath the selected function. A new blade will open up which allows a user to update the trigger and add any bindings. When using table storage, it needs the connection string to the appropriate storage account, the name of the table where the data is stored, and a parameter name. There are other fields to fill out, but they are optional and not required for the experiment. The connection string can be updated in the storage account connection field option. Clicking the *new* option next to the field reveals all storage accounts associated with the current subscription. Selecting the correct storage account will automatically import the connection string. When inputting the table, only the table name is required. Unlike blob storage, table storage does not allow sub containers, so all tables are stored in the same location under one storage account [63].

The parameter name is the name used in the function when referencing the actual table. For example, the experiment's table name was battery5TableStorage, and the parameter name was inputTable. Any time a read operation was performed on the battery5TableStorage table in the function, the inputTable parameter name would be used. The documentation online on how to perform this basic setup is lacking. While intuitive in hindsight, it was initially very confusing. After updating the bindings through the portal, the existing JSON file should update automatically to reflect the changes made.

Once the bindings are updated, modifications can be made to the function code. The parameter names used in the JSON file must be used as actual parameter names for the function. The datatype for these parameters depends on which function runtime is being used. The two versions are 1.x, for legacy runtimes, and 2.x, for newer runtimes [67]. If using 1.x, the datatype used is called IQueryable. If using 2. x, the datatype used is called CloudTable. Developing functions in the portal automatically uses version 2.x, so each parameter representing a table should be a CloudTable datatype. Continuing the example from above, the full function parameter name would be CloudTable inputTable.

After the function parameters are declared, it would be nice if the process was over and a user could simply use the CloudTable parameters directly. But there is still more setup required. To get information from the table, it can be queried using TableQuery and TableQuerySegment datatypes [68]. A query is provided to a TableQuery, which in turn is passed to a TableQuerySegment to get information from the table. After the results are returned from the query, one final step remains. A class must be declared whose instance variables match the names of the columns in the table to be read. The user may select which columns to use as instance variables for the class, but the partition key and row key are included by default. Take note of the datatypes of each column; the datatype of each instance variable must match the datatype of the corresponding column. If unsure of the datatypes, use the Microsoft Azure Storage Explorer to view specific details about each column. Once the class is created, data can finally be processed and analyzed by iterating over each row in the results sequentially. Each row data is stored in an object of the class's instance variables allowing for easy access and modification of the data.

To store data to a table from a function, the output binding must be added to the JSON file similar to how input bindings were added. The chosen parameter name must be added to the function parameter list as well. For example, if an output table was declared in the bindings called outputTable, the full function parameter name would be CloudTable outputTable. No querying is necessary to write to the output table, so the TableQuery and TableQuerySegment datatypes are not needed. Instead, a TableOperation datatype is used [68]. An object must be created which stores any relevant data for a row to be written. This object is passed to the TableOperation which then writes the data to storage. Online documentation shows other ways to read and write from table storage, but the examples are limited and don't provide much detail if an error occurs. Figures 9.21 and 9.22 show the function used to read battery data from one table, perform some computations on the data, and then output the results to a different table.

9.3.7 Azure Machine Learning

When the telemetry data has been processed and stored, a machine learning model can analyze the processed data to perform predictive analytics. The process of creating a predictive model is straightforward. First data must be collected and processed, where processing means transforming the data into a more readable format for a machine learning algorithm to analyze. Then an algorithm must be selected to act as the predictive backbone for the model. Data is then passed to the selected algorithm to train and create a model. Once the model is trained and fully tested, it can be deployed and used for real-time prediction.

When selecting an algorithm, there are three main categories, supervised, unsupervised, and reinforcement learning [69, 70]. Supervised learning uses clearly labeled historical data to train a model and provide predictions. The equation $f(x) = y$ is a simple way to explain how it works. The function f, the model, accepts the variable x, the historical input data often referred to as the features, as a parameter which computes y, the variable to predict. Typically the input data comes in a table format, where each row represents a single entity and each column is an attribute for each entity. For example, a table could contain information about a list of houses. Each row is a different house, and the columns describe the number of the bathrooms, bedrooms, square footage, and so on for each house.

Supervised learning can be separated into classification and regression, with each category offering a different type of prediction. Classification algorithms focus on creating groups with similar traits, where prediction is used to classify new data as belonging to one of these groups. Classification is well suited for spam detection, where a model is trained using millions of different emails where each email is labeled as spam or not spam. When the model is trained, new emails that are not labeled can be passed to the model to predict the group it belongs to. Regression algorithms are used to predict future values, typically numerical. A common example for regression is predicting housing prices. Feature selection can include many of

```
 1  #r "Newtonsoft.Json"
 2  #r "Microsoft.WindowsAzure.Storage"
 3
 4  using Microsoft.WindowsAzure.Storage.Table;
 5  using Microsoft.WindowsAzure.Storage;
 6  using System.Threading.Tasks;
 7  using Microsoft.Extensions.Logging;
 8  using Microsoft.AspNetCore.Mvc;
 9  using Microsoft.Extensions.Primitives;
10  using Newtonsoft.Json;
11  using System.Net;
12
13  public static async void Run(HttpRequest req, CloudTable inputTable, CloudTable outputTable, ILogger log)
14  {
15      //TableQuery used to store query.
16      TableQuery<ReadEntity> query = new TableQuery<ReadEntity>().
17          Where(TableQuery.GenerateFilterCondition("Batt_name", QueryComparisons.Equal, "B0005"));
18      TableContinuationToken token = null;
19      List<ReadEntity> allEntities = new List<ReadEntity>();
20
21      do
22      {
23          //TableQuerySegment uses TableQuery to search inputTable.
24          TableQuerySegment<ReadEntity> resultSegment =
25              await inputTable.ExecuteQuerySegmentedAsync(query, token);
26          //Token used to continue searching large tables.
27          //If not used, TableQuerySegment will only return 1000 rows.
28          token = resultSegment.ContinuationToken;
29
30          //InputTable may not be sorted.
31          //Must store in list to sort before performing computations.
32          foreach (ReadEntity entity in resultSegment.Results)
33          {
34              allEntities.Add(entity);
35          }
36      } while (token != null);
37
38      allEntities.Sort(new TableSorter());
39
40      double coulomb = 0;
41      bool hasWritten = false;
42      ReadEntity prev,cur;
43
44      for(int i = 1; i < allEntities.Count; i++)
45      {
46          prev = allEntities[i-1];
47          cur = allEntities[i];
48              if(prev.cycle == cur.cycle)
49              {
50                  if(prev.voltage_battery < 2.7)
51                  {
52                      //Add coulomb data to current object and write to table.
53                      prev.coulomb = coulomb/3600;
54                      TableOperation insertOperation = TableOperation.Insert(prev);
55                      await outputTable.ExecuteAsync(insertOperation);
56                      hasWritten = true;
57                  }
58                  else if(!hasWritten)
59                  {
60                      //Compute coulomb.
61                      coulomb += (Convert.ToDouble(cur.time) - Convert.ToDouble(prev.time)) *
62                          Convert.ToDouble(cur.current_battery);
63                  }
64              }
65              else
66              {
67                  //New cycle has started. Reset Values.
68                  coulomb = 0;
69                  hasWritten = false;
70              }
71      }
72      log.LogInformation("Ending process.");
73  }
```

Fig. 9.21 Azure Function used to read and write to table storage. See Fig. 9.22 for classes required to complete function

```
75 //Class to store input data from table storage.
76 public class ReadEntity : TableEntity
77 {
78      public string Batt_name { get; set;}
79      public string dateTime { get; set;}
80      public long MessageID { get; set;}
81      public long cycle { get; set;}
82      public double time { get; set;}
83      public double voltage_battery { get; set;}
84      public double voltage_load { get; set;}
85      public double current_battery { get; set;}
86      public double current_load{ get; set;}
87      public double temp_battery { get; set;}
88      //Variable not part of input table.
89      //Added as new compute variable to help prediction.
90      public double capacity { get; set; }
91 }
92
93 //Class to sort all inputs by messageID.
94 class TableSorter : IComparer<ReadEntity>
95 {
96      public int Compare(ReadEntity x, ReadEntity y)
97      {
98          return (int)(x.MessageID - y.MessageID);
99      }
100 }
```

Fig. 9.22 Classes used to help Azure Function read and write to table storage. See Fig. 9.21 for main method of Azure Function

the attributes listed previously and any other relevant information. The model is trained using these features along with the houses' current value. Once trained, data from new houses can be input to the model to predict its future price. For the battery experiment, some form of regression would be the best choice for predicting the cycle, after which the battery will need to be replaced.

While unsupervised and reinforcement learning were not used, they will be discussed briefly here to complete the machine learning picture. Unsupervised learning is used for learning without knowing the desired outcome. It is more of an exploratory process where there is no specific output value to predict unlike supervised learning. Some uses are clustering, association, and anomaly detection. Clustering and association find hidden relations among the data. Clustering is like supervised classification where similar or dissimilar data is grouped together, but no predictions are made. It is just grouping the data together based on relations. Advertising firms can use clustering to group populations to better understand and target an area for a product. Association can be used by retailers to find correlation between buying certain products to help increase sales. Anomaly detection is used for finding unexpected behavior. Credit card companies can use this for fraud detection, where unexplained spending can be detected and the customer can be alerted. Reinforcement learning involves a model being able to learn in an environment where rewards and punishments are given based on its actions. The goal is to

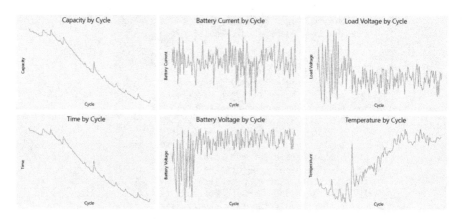

Fig. 9.23 Plotting relations between each feature and cycle in *Power BI Desktop*. Battery current, battery voltage, load voltage, and temperature are all instances of the data set when the load voltage dropped below 2.7. Capacity and time are computed from the start of each cycle to the point where load voltage drops below 2.7

maximize the rewards while minimizing punishments. The company DeepMind Technologies is famous for creating AlphaZero, a program that uses reinforcement learning to play the games of chess, shogi, and go at superhuman levels [71].

When selecting features to train the model, it is a common practice to visualize the data before training a model. Visualizing the battery experiment data, where the x-coordinates are one feature and the y-coordinates are another, can reveal several relations between different features helping with feature selection. See Fig. 9.23 for an example, which shows the battery data imported and visualized with different plots in *Power BI Desktop*. The cycle of the battery will be a feature, so comparing other features to it may reveal significant trends. Capacity and time, where time is the length of each cycle, have the strongest relation to a cycle, so they should be included in the feature selection. Temperature also has some relation, so it may also prove useful. The rest are more erratic with no real trends, which means they may not be the best choices to include.

The visuals can also be used to help with algorithm selection. Many of the less erratic visuals have a sloping trend, so linear or polynomial regression could be good choices. Linear regression tries to find a straight line that closely represents the general trend of the data, and polynomial regression creates a similar line except the line can curve to follow the data, like a polynomial function. Trial and error will always be a part of the process, where many features should be tested against many different algorithms. Using visuals can help narrow down the best features and algorithms to choose from.

It is straightforward to understanding the process of creating a model. However, it is not easy to collect and process raw data, to select features, to choose an algorithm, and to write a program that combines all these parts successfully. Luckily Azure provides services for all skill levels, where even users new to machine learning can create models in minutes. Selecting one of these services can be confusing though, as there are cloud-based, on-premise, and development platforms and tool options. Also, some of the names used for different services are very similar, which can add to the confusion. The cloud-based options are *Azure Machine Learning Studio* and *Azure Machine Learning Services*. On-premise options are *SQL Server Machine Learning Services* and *Microsoft Machine Learning Server*. The development platforms and tools are *Azure Data Science Virtual Machine*, *Azure Databricks*, *ML. NET*, and *Windows ML* [72]. The cloud-based options will be the focus for this section, with an emphasis on *Azure Machine Learning Studio* as it was used for the battery experiment. *Azure Machine Learning Studio* is great for creating, training, and deploying models quickly with ease. It is also better for smaller projects, where scaling the model to accept larger amounts of data is not an issue.

Azure Machine Learning Studio is a drag and drop interface which focuses on placing and connecting modules to build a model [73]. It does not run in the *Azure portal*, but can be accessed through the *Azure portal* using a *Machine Learning Studio workspaces* resource or via normal website addressing. A similar version is available in the portal called *visual interface*, but it is accessible using a *Machine Learning service workspaces* resource which will be discussed later. The layout of the interface is very intuitive and easy to understand, with several tabs to make organization and creating models easier. The projects tab is used for creating projects, which act as containers for organizing experiments and files. The experiments tab is where the design and creation of a model begins. Take note that models are called training or predictive experiments in studio. The web services tab contains all web services. A web service is a fully trained predictive experiment that is accessible by other users and programs.

A free and standard version of *Azure Machine Learning Studio* is available where there are similarities and differences between both. Both versions can create predictive models and deploy classic web services. The standard version does have a monthly subscription fee, but it offers more storage, allows for larger models to be built, can train models faster, has a service-level agreement, and can deploy new web services which offer more features compared to classic web services. If using the standard version, a *Machine Learning Studio workspaces* resource must be created in the portal so compute and storage accounts can be allocated. The free version does not require a workspaces resource or need any connection to the portal.

To create a new experiment, select the experiments tab, and click the new button to bring up a list of options. The user can create a blank experiment or use one of the many examples Microsoft provides as a starting point or guide to creating their own [74]. Selecting a blank experiment opens a new workspace where modules can now be selected, dragged, and dropped into. After dropping a few modules in the experiment, they can be connected with lines denoting the control flow of the experiment.

Every module is self-contained meaning no code is required to use it [75]. Each module can be customized to some extent, but most of the implementation is hidden from the user. Each module performs some task which sends its results to the next module, and the names of the modules typically describe their task. For example, a module titled Import Data is used for importing a data set. This module can connect to Azure storage, but the users' subscription level may limit this feature [76]. If this is the case, the data can be uploaded directly to the workspace. There are a wide array of modules to select from whose operations range from data transformation, statistical analysis, performing SQL commands, machine learning algorithms, and many more. Of the machine learning algorithms, regression, classification, anomaly detection, and clustering algorithms are all available with each category offering multiple choices. If a module does not exist for a certain task, the Execute Python Script and Execute R Script modules are available to implement custom code. This allows users to perform custom data transformations, create and use different metrics for analysis, and even run different algorithms not available as an existing module.

Once the data is imported and features are selected, one way to create a model is to use the Split Data, Train Model, Score Model, and Evaluate Model modules. The Split Data module takes data as input and splits the data into training and testing sets. The training set is used to train the model, where the testing set is used to see how accurate the model is once trained. By default, 70% of the data becomes training, where 30% becomes testing, but this number can be changed by the user. Randomization can also be applied when splitting the data, and it is highly recommended to do so. The training set is used as input for the Train Model module, along with the output of the module for the algorithm chosen. For the battery experiment, the *Linear Regression* module was chosen as the algorithm.

The Train Model module trains and creates a model for prediction. Each algorithm module has different hyperparameters that can be adjusted to train the model differently. Feature selection and model selection are extremely important, as changing hyperparameters and selecting different features can increase or decrease prediction accuracy. Visualizing the data before selection and training is one way to help select both as described earlier. The output model is combined with the testing set in the Score Model module to compute predictions for the testing set. This module does not give an overall score for the model, but each individual prediction

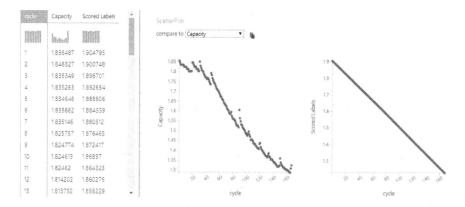

Fig. 9.24 Some visualization features comparing real to predicted values

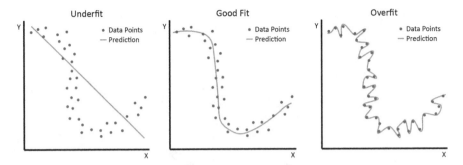

Fig. 9.25 Underfitting, good fitting, and overfitting examples for machine learning [77]

can be compared to its actual value. See Fig. 9.24 for an example comparing the two values along with some other visualization features offered in studio.

To get the overall accuracy of the model, the output from the score module can be sent to the Evaluate Model module. Good accuracy scores can be anything above 75%, but there is no set number for a good score as it may vary based on the complexity of the model and the predictions being made. If a number is too high, say 99% or 100%, the model is probably overfitted. This means too much noise is being captured when training which causes the model to fit the training data too well. If the number is very low, the model is probably underfitted. This occurs when no general trends are found in the data and the model is oversimplified [77]. If either occurs, adding or removing features from the data, modifying the hyperparameters of the current model, and selecting a new algorithm are possible fixes. Both scenarios can lead to very poor predictions of new data if not addressed. See Fig. 9.25 for examples of underfitting, a good fit, and overfitting.

Another way to create a model is to use the *Cross Validate Model* module. It splits, trains, scores, and evaluates the model all in one module. It uses a technique called cross-validation to split the data randomly into equal folds, or partitions [78]. The default number of folds is ten, but this can be changed using the Partition and Sample module if desired. The model is trained using all but one of the folds. The fold not used, called the holdout fold, is used for testing the trained model. This process is repeated, so each fold is used as testing data. For example, if there are ten folds, the model would be trained ten times. During each training iteration, a different holdout fold is selected where the remaining nine folds are used for training. Evaluation is done during each iteration and stored; if iteration scores are vastly different, then there may be problems with the model or data. The advantages of using cross-validation over splitting the data are as follows: it uses more data for training and testing making it better for smaller data sets, it can help prevent overfitting, and it evaluates the quality of the data being used as well. A disadvantage is it is more computationally expensive, so larger data sets may take more time. See Fig. 9.26 for an example of a battery training experiment using both methods described to create a model.

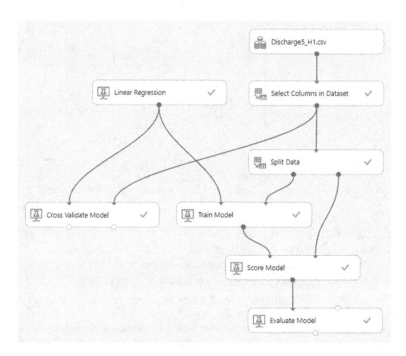

Fig. 9.26 Experiment using two different methods to train a model

When a user thinks the training experiments prediction capabilities are sufficient, it must be deployed as a web service so it may be used against new data. The process is very easy: click the Set Up Web Service option, and then select Predictive Web Service. This will automatically start shuffling, adding, and removing modules to convert the training experiment to a predictive experiment. If multiple algorithm modules are being tested in one experiment, it is recommended to remove all unwanted modules unrelated to the selected algorithm, so only the relevant modules remain [79]. The conversion process will not remove the module used to import data, as it is used as a template for what features will be passed into the predictive experiment. When building the training experiment, the variable to predict was allowed to pass through the Select Columns in Dataset module as it was needed for training the model; with new data this value is unknown, so it can be removed from feature selection in the predictive experiment. See Fig. 9.27 for the result of converting Fig. 9.26 to a predictive experiment.

When complete, the Deploy as Web Service option can be selected which will add the model to the web services tab as a classic or new web service. As mentioned previously, the free version of studio can only create classic web services, where the standard version can create both. From the web services tab in *Azure Machine Learning Studio*, the newly created web service should appear and can now be used for prediction. Single values or batch testing can be done through the browser using *Azure Machine Learning Web Services* or Excel. *Azure Machine Learning*

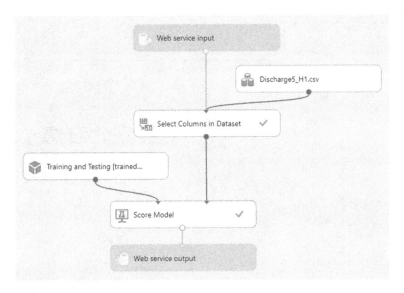

Fig. 9.27 Result of converting Fig. 9.26 experiment to a web service prediction model

Web Services is another option for managing, testing, and deleting web services created in studio [80]. Selecting the *New Web Services Experience* link from the web service tab in studio will open up *Azure Machine Learning Web Services* in a new web page. The *Test* tab contains all options for running predictions against the model.

If using Excel, the add-in for Azure Machine Learning must be installed, and the URL and API key must be entered. The URL and API keys are found in the *Consume* tab from the *Azure Machine Learning Web Services* page. The user may also click the Excel link from the studio or web services page, and this will download a blank Excel file with the Azure Machine Learning add-in, URL, and API keys installed. Data in Excel can be selected and sent to the web service for prediction, with the results being returned and populated in Excel.

Azure Machine Learning Services is another cloud-based machine learning platform in Azure. It offers multiple ways to build, train, and manage predictive models from multiple sources in the cloud. It is *Azure portal* based, meaning a user must create a *Machine Learning Service Workspaces* resource to use it. Predictive models can be created using *visual interface*, *automated machine learning*, or the *Azure Machine Learning Python SDK* which can be used with multiple programming interfaces [81]. Currently several of the tools are in preview stage, which means the service should primarily be used for testing or proof of concept. Services labeled as preview can be removed at any time, have different service-level agreements, and are used as is without expectations of further updates [82].

Visual interface is a partial clone of *Azure Machine Learning Studio* with some minor differences. *Visual interface* is a drag and drop workspace to create and train models using prebuilt modules, but everything is done in the *Azure portal*, so it is

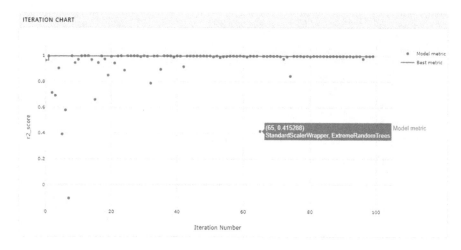

Fig. 9.28 Automated machine learning results

fully integrated with *Azure Machine Learning Services* unlike studio [83]. It also has fewer modules to choose from, and it relies on a user's own compute resources to train and test models [73]. Compute resources are created by the user, used by the workspace to run and train models, and can be located on a local machine or in the cloud. Like other Azure services, compute resources running in the cloud are only billed when they are active. When training and testing, it is recommended to use a local machine with smaller data sets. When the model is ready, compute resources can be pushed to the cloud where they can scale up or down based on demand.

Automated machine learning, or sometimes referred to as *automated ML*, is a way for Azure to train, test, and tune multiple models at a time [84]. The user only needs to provide a data set, feature selection, prediction variable, and whether they want to perform regression, classification, or forecasting. When automation is started, Azure will train many models using multiple algorithms, modifying the hyperparameters during each iteration in an attempt to create the best model for prediction. When complete, all the models trained are presented to the user with prediction scores and other evaluation metrics. *Automated ML* can be run in the portal or by using the *Azure Machine Learning Python SDK* in Python. Figure 9.28 shows a sample run using the *Azure portal* on the battery data set. Each blue dot is not a different model but a model's score during one of its training iterations. The sample run took about 15 min. and created 26 different models. Many of the scores are extremely high though, so there is a good chance many of the models are overfitted.

If a user prefers more control and writing their own code, the *Azure Machine Learning Python SDK* can be used with *Jupyter Notebook*, any Python IDE, or Visual Studio Code to connect to *Azure Machine Learning Services* [85]. *Jupyter Notebook* is an open-source web application that allows users to write and share code through a browser [86]. Microsoft also offers a service called *Azure Notebooks*, which essentially is Jupyter Notebook, but it runs in the cloud and requires no

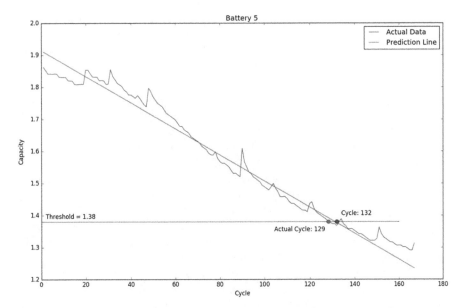

Fig. 9.29 Output created using python script module to show prediction

installation [87]. By importing the SDK, any Python environment can manage data sets or cloud resources, train models, deploy web services, and use *automated machine learning*. Using the SDK requires more knowledge about programming, where *automated ML* and *visual interface* are code free.

Many of the steps used to create a predictive model for the battery experiment have been detailed in the previous section focusing on *Azure Machine Learning Studio*, even though many of the tools in *Azure Machine Learning Services* could have been used as well. Figure 9.29 shows a plot using the *Execute Python Script* module comparing the training data to the prediction line. For this data set, the predicted replacement cycle is only off by three compared to when the battery should be replaced.

9.3.8 Power BI

Many of Azure services provide some basic visualization options for data analysis, but they are limited in terms of customization and creating presentations. Microsoft offers *Power BI*, a collaborative drag and drop visualization tool that allows users to transform their data into interactive presentations. There are several services offered in *Power BI*, but three useful tools for creating presentations and delivering action-able business insights are *Power BI Desktop*, *Power BI service*, and *Power BI mobile apps* [88].

Power BI Desktop is a free, downloadable software that allows users to connect, import, modify, and visualize their data on their own computer [21]. When importing or connecting to the data source, the *Query Editor* is used to remove columns, change datatypes, or perform other modifications to the data before creating visuals. Modifications made are saved in the query, so each time the file is opened again, the data remains in the same format. This also keeps the source data intact and allows queries to be reversed [89]. Special attention should be placed on the column datatype, as datatypes determine how the data is visualized. When the data is in the correct format, the *Query Editor* can be closed, and the data can be used to create visualizations. Visualizations, sometimes called visuals, are pre-configured templates for bar charts, line charts, scatter charts, maps, and many other options. Each visualization has options that can be further edited to fit a specific visual style, such as modifying colors, adding text, and selecting a range for the data. Visuals can be placed on report pages, which are canvases to drag and drop visuals onto. Typically *Power BI Desktop* is used for creating reports, and not sharing them. To share reports, they can be published and uploaded to *Power BI service*.

Power BI service is a cloud-based, Software as a Service (SaaS) tool primarily used for collaboration on pre-built models from *Power BI Desktop*. Both can create and edit reports, but *Power BI Desktop* is better suited for it [90]. The key components of a *Power BI service* are workspaces, dashboards, reports, data sets, and workbooks [91]. Workspaces are containers for the other key components of a *Power BI service*. Workspaces can either be a my workspace or app workspaces. A my workspace is a personal workspace for a single user, where only they can access and modify its content. App workspaces are for collaboration and sharing workspace content with other users. They also are used for app creation where end users can interact with the data but not modify it. Dashboards are a single, shareable canvas that can display visualizations from multiple reports, data sets, and other tools such as Excel. They are useful for grouping many related visualizations together to aid in business decisions. Reports consist of one or more pages of visualizations in a workspace. A report can only get data from one data set, but reports can belong to multiple dashboards. Data sets are data a user wishes to visualize. They can be imported or connected to and be used in multiple workspaces. Workbooks are similar to data sets, except the data is from an Excel workbook.

Embedded analytics is the process of inserting meaningful visuals, reports, and dashboards into an application through application program interfaces (APIs) to give a customer or an organization business insights for better decision-making. Embedding can be done using the APIs in *Power BI service*, or through the *Azure portal* with *Power BI Embedded service* [92]. Embedding for a customer and embedding for an organization are two different tasks. Embedding for a customer allows a client to access visuals, reports, and dashboards without having a *Power BI* account. The customer does not even need to know how to use *Power BI*, as much of the implementation is hidden from them. Embedding for an organization requires all users of the organization to have a *Power BI* account. Each user is given access to information only required for their work, narrowing the information flow and giving the user more focus to relevant information for their task. Another way to view

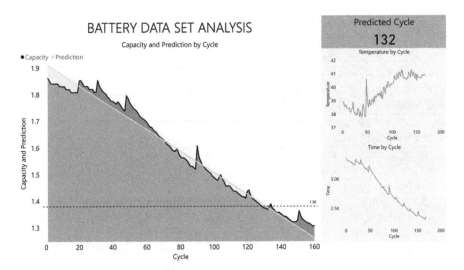

Fig. 9.30   Power BI report showing prediction compared to actual values. Temperature (in Celsius) is the instance of the data set when the battery voltage dropped below 2.7. Capacity (in Coulomb) and time (in seconds) are computed from the start of each cycle to the point where battery voltage drops below 2.7

information is through *Power BI mobile apps*. The apps are available for Apple, Android, and Windows 10 devices. Users simply download the app and sign in. The app will list all dashboards and reports available so they can be monitored from any device. Dashboards and reports cannot be created using *Power BI mobile apps*, which must be done in *Power BI service* or *Power BI Desktop*.

For the battery experiment, an Excel file with actual and predicted data was loaded into *Power BI Desktop*. After the file is loaded, the *Query Editor* will open allowing for the modification of the data. When complete, clicking "close and apply" will cause the Query Editor to exit after applying all changes to the data set. With the data loaded, visualizations can be added. Clicking any of the options in the Visualizations tab will cause a generic box to appear in the report page. With the generic box selected, columns can be dragged and dropped into the settings of the box, which are below the Visualizations tab. After a few columns are added, the box will populate itself with the corresponding values. Visualizations can be changed to any other visualization at any time. If a visual is a stacked bar chart, selecting stacked area chart will update the visual automatically. Some visualizations may not have a one-to-one correlation, such as a line chart and a pie chart, so some of the settings will need to be modified. The whole process is very intuitive, and as it is a design process, trial and error will be involved. See Fig. 9.30 for example output from the data set.

9.3.9 Experiment Observations

Throughout this chapter, many of the steps used to create the experiment have been described and shown in great detail. After all the setup is complete and the predictive model is released as a web service, new batteries can be connected to sensors which will send their messages to the *IoT Hub*. *Stream Analytics* can monitor these messages and store the data in a table while scanning the current battery cycle. When the battery voltage dips below 2.7 for the specified cycle, it will trigger a *Azure Function* to read all relevant data for the new battery and compute the necessary values. The computed values can be sent to the predictive web service to get a better idea when the battery should be changed out.

If a user wants to cut costs or to ensure that the amount of data sent to the IoT hub is manageable for human observers, the user can create a custom endpoint for the *IoT Hub* resource, which stores all telemetry data to blob storage. The user can then monitor the cycles on their own and wait till the desired cycle is reached to launch a slightly modified *Azure Function* to read from blob storage and to compute the new values. This eliminates *Stream Analytics* from the process, which can greatly reduce the overall costs. For the battery experiment, it is possible to do this as the time intervals between the MQTT messages are large enough for a human observer to manage. There will be some loss in efficiency, as *Stream Analytics* triggers the *Azure Function* the moment it finds the value where a human will have to continuously check. A hybrid approach is also possible, where telemetry data sent to the *IoT Hub* is stored in blob storage and *Stream Analytics* is started much later in the batteries' life cycle. This reduces the overall costs and maintains efficiency. If multiple batteries are being monitored or the time intervals decreased to a speed humans cannot perceive, *Stream Analytics* should be used, and the costs are unavoidable.

9.4 Conclusion

Developing an IoT applications on a cloud platform requires thorough understanding of the programming API, configuration, performance, limitation, and design patterns of the tools of the platform. In this chapter, we reviewed the tools of Azure IoT platform including device clients, gateways such as *event hub*, IoT communication protocols, cloud storage such as Cosmos DB and *blob storage*, analysis tools such as *Stream Analytics* and *Azure Functions*, and visualization tools such as Power BI. We did not discuss Azure components related to monitoring, logging, backend application integration, and machine learning since they are highly related to specific IoT applications. IoT platform development is still research in progress, and new tools are emerging every year, and the API and performance of the existing tools are improving rapidly. While Azure IoT platform is already a scalable, extensible, and comprehensive development environment, it still has many limitations that may be addressed in its future offerings. For example, Azure IoT platform is not suitable for applications with low-latency requirements. These applications need to either have on-premise gateway, data analysis, and visualization components or build customized cloud IoT system that can satisfy latency constraints. Specific IoT applications may use selected parts of the Azure IoT ecosystem while replying on customized components to satisfy performance requirements. For example, Power BI is not suitable for real-time streaming with high ingestion rate. Users can develop customized visualization tool using low-level libraries such as D3.js to render high-frequency data retrieved from Azure *IoT hub*.

References

1. Jay Lee, Hung-An Kao, Shanhu Yang, Service innovation and smart analytics for industry 4.0 and big data environment. Procedia CIRP, 16:3–8, 2014. Product Services Systems and Value Creation. Proceedings of the 6th CIRP conference on industrial product-service systems
2. D. Sonntag, S. Zillner, P. van der Smagt, A. L¨orincz, *Overview of the CPS for Smart Factories Project: Deep Learning, Knowledge Acquisition, Anomaly Detection and Intelligent User Interfaces* (Springer International Publishing, Cham, 2017), pp. 487–504
3. D. Alahakoon, X. Yu, Smart electricity meter data intelligence for future energy systems: A survey. IEEE Trans. Ind. Inf. **12**(1), 425–436 (2016)
4. S. Jain, , A. Paventhan, V. Kumar Chinnaiyan, V. Arnachalam Survey on smart grid technologies- smart metering, iot and ems. In 2014 IEEE Students' Conference on Electrical, Electronics and Computer Science, 1–6, 2014, March
5. F.J. Valente, A.C. Neto, Intelligent steel inventory tracking with iot / rfid, in *2017 IEEE International Conference on RFID Technology Application (RFID-TA)*, (2017, Sep), pp. 158–163
6. R. Vargheese, H. Dahir, An iot/ioe enabled architecture framework for precision on shelf availability: Enhancing proactive shopper experience, in *2014 IEEE International Conference on Big Data (Big Data)*, (2014, Oct), pp. 21–26
7. M. Bacco, A. Berton, E. Ferro, C. Gennaro, A. Gotta, S. Matteoli, F. Paonessa, M. Ruggeri, G. Virone, A. Zanella, Smart farming: Opportunities, challenges and technology enablers, in *2018 IoT Vertical and Topical Summit on Agriculture - Tuscany (IOT Tuscany)*, (2018, May), pp. 1–6
8. W. He, G. Yan, L.D. Xu, Developing vehicular data cloud services in the iot environment. IEEE Trans. Ind. Inf. **10**(2), 1587–1595 (2014)
9. B. Padmaja, P.V. Narasimha Rao, M. Madhu Bala, E.K. Rao Patro, A novel design of autonomous cars using iot and visual features, in *2018 2nd International Conference on I-SMAC (IoT in Social, Mobile, Analytics and Cloud) (I-SMAC)I-SMAC (IoT in Social, Mobile, Analytics and Cloud) (I-SMAC), 2018 2nd International Conference on*, (2018, Aug), pp. 18–21
10. J. Backman, J. Väre, K. Främling, M. Madhikermi, O. Nykänen, Iot-based inter- operability framework for asset and fleet management, in *2016 IEEE 21st International Conference on Emerging Technologies and Factory Automation (ETFA)*, (2016, Sep), pp. 1–4
11. H. Hejazi, H. Rajab, T. Cinkler, L. Lengyel, Survey of platforms for massive iot, in *2018 IEEE International Conference on Future IoT Technologies (Future IoT)*, (2018, Jan), pp. 1–8
12. P. Ray, A survey of iot cloud platforms. Futur. Comput. Inform. J **1**(1–2), 35–46 (2016)
13. T. Pflanzner, A. Kertesz, A survey of iot cloud providers, in *2016 39th International Convention on Information and Communication Technology, Electronics and Microelectronics (MIPRO)*, (2016, May), pp. 730–735

14. D. Clerissi, M. Leotta, G. Reggio, F. Ricca, Towards an ap- proach for developing and testing node-red iot systems, in *Proceedings of the 1st ACM SIGSOFT International Workshop on Ensemble-Based Software Engineering* . EnSEm- ble 2018, (ACM, New York, 2018), pp. 1–8
15. A.A. Ismail, H.S. Hamza, A.M. Kotb, Performance evaluation of open source iot platforms, in *2018 IEEE Global Conference on Internet of Things (GCIoT)*, (2018, Dec), pp. 1–5
16. M. Moravcik, P. Segec, M. Kontsek, Overview of cloud computing standards, in *2018 16th International Conference on Emerging eLearning Technologies and Applications (ICETA)*, (2018, Nov), pp. 395–402
17. P. Garćia L'opez, M. S'anchez-Artigas, G. Pańis, D. Barcelona Pons, A. Ruiz Ollobarren, D. Arroyo Pinto, Comparison of faas orchestration systems. in *2018 IEEE/ACM International Conference on Utility and Cloud Computing Companion (UCC Compan- ion)*, pp. 148–153 (2018, Dec)
18. J. Gibson, R. Rondeau, D. Eveleigh, Q. Tan, Benefits and challenges of three cloud computing service models, in *2012 Fourth International Conference on Computational Aspects of Social Networks (CASoN)*, (2012, Nov), pp. 198–205
19. Microsoft, What is Azure IoT Hub? (2018), https://docs.microsoft.com/en-us/azure/iot-hub/about-iot-hub. Accessed 11 April 2019
20. Microsoft, Stream analytics documentation (2019), https://docs.microsoft.com/en-us/azure/stream-analytics/. Accessed 11 April 2019
21. Microsoft, What is Power BI Desktop? (2019), https://docs.microsoft.com/en-us/power-bi/desktop-what-is-desktop. Accessed 11 April 2019
22. Microsoft, What is Azure Blob storage? (2018), https://docs.microsoft.com/en-us/azure/storage/blobs/storage-blobs-overview. Accessed 11 April 2019
23. Microsoft, Welcome to Azure Cosmos DB (2019), https://docs.microsoft.com/en-us/azure/cosmos-db/introduction. Accessed 11 April 2019
24. Microsoft, Azure Functions (2017), https://docs.microsoft.com/en-us/azure/azure-functions. Accessed 10 April 2019
25. K. Goyal, A. Garg, A. Rastogi, S. Singhal, A literature survey on internet of things (iot). Int. J. Adv. Netw. Appl **9**, 3663–3668 (2018)
26. A. Polianytsia, O. Starkova, K. Herasymenko, Survey of the iot data transmission protocols, in *2017 4th International Scientific-Practical Conference Problems of Infocommunications. Science and Technology (PIC S T)*, (2017, Oct), pp. 369–371
27. Oasis. MQTT Version 3.1.1 (2014), http://docs.oasis-open.org/mqtt/mqtt/v3.1.1/os/mqtt-v3.1.1-os.html. Accessed 11 April 2019
28. N. Naik, Choice of effective messaging protocols for iot systems: Mqtt, coap, amqp and http, in *2017 IEEE International Systems Engineering Symposium (ISSE)*, (2017, Oct), pp. 1–7
29. Microsoft, Azure IoT SDK (2018), https://docs.microsoft.com/en-us/azure/iot-hub/iot-hub-devguide-sdks. Accessed 10 April 2019
30. Microsoft, Azure IoT protocol gateway (2017), https://docs.microsoft.com/en-us/azure/iot-hub/iot-hub-protocol-gateway. Accessed 10 April 2019
31. Microsoft, Azure Event Hubs — A big data streaming platform and event ingestion service? (2018), https://docs.microsoft.com/en-us/azure/event-hubs/event-hubs-about. Accessed 11 April 2019
32. Microsoft, Security for Internet of Things (IoT) from the ground up (2018), https://docs.microsoft.com/en-us/azure/iot-fundamentals/iot-security-ground-up?context=azure/iot-hub/rc/rc. Accessed 11 April 2019
33. Microsoft, Provisioning devices with Azure IoT Hub Device Provisioning Service (2019), https://docs.microsoft.com/en-us/azure/iot-dps/about-iot-dps#when-to-use-device-provisioning-service. Accessed 11 April 2019
34. Microsoft, Azure Blob storage: hot, cool, and archive access tiers (2019), https://docs.microsoft.com/en-us/azure/storage/blobs/storage-blob-storage-tiers. Accessed 11 April 2019
35. Microsoft, What is Azure SQL Database service (2019), https://docs.microsoft.com/en-us/azure/sql-database/sql-database-technical-overview. Accessed 11 April 2019

36. Microsoft, Introduction to Azure Data Lake Storage Gen2 (2018), https://docs.microsoft.com/en-us/azure/storage/blobs/data-lake-storage-introduction. Accessed 11 April 2019
37. E. Siow, T. Tiropanis, W. Hall, Analytics for the internet of things: A survey. ACM Comput. Surv. **51**(4), 74:1–74:36 (2018)
38. S. Singh, Optimize cloud computations using edge computing, in *2017 International Conference on Big Data, IoT and Data Science (BID)*, (2017,Dec), pp. 49–53
39. N.K. Giang, R. Lea, M. Blackstock, V.C.M. Leung, Fog at the edge: Experiences building an edge computing platform, in *2018 IEEE International Conference on Edge Computing (EDGE)*, (2018, July), pp. 9–16
40. Microsoft, Stream analytics query language reference (2016), https://docs.microsoft.com/en-us/stream-analytics-query/stream-analytics-query-language-reference. Accessed 10 April 2019
41. Microsoft, Introduction to stream analytics windowing functions (2019), https://docs.microsoft.com/en-us/azure/stream-analytics/stream-analytics-window-functions. Accessed 11 April 2019
42. Microsoft, Azure Stream analytics – user defined functions (2018), https://docs.microsoft.com/en-us/azure/stream-analytics/stream-analytics-javascript-user-defined-functions. Accessed 10 April 2019
43. Microsoft, Azure Stream analytics – user defined aggregates (2017), https://docs.microsoft.com/en-us/azure/stream-analytics/stream-analytics-javascript-user-defined-aggregates. Accessed 10 April 2019
44. Microsoft, Run Azure Functions from Azure Stream Analytics jobs (2019), https://github.com/MicrosoftDocs/azure-docs/blob/master/articles/stream-analytics/stream-analytics-with-azure-functions.md. Accessed 10 April 2019
45. Microsoft, Durable Azure Functions (2018), https://docs.microsoft.com/en-us/dotnet/standard/serverless-architecture/durable-azure-functions. Accessed 10 April 2019
46. Microsoft, Processing 100,000 events per second on Azure Functions (2017), https://azure.microsoft.com/en-us/blog/processing-100-000-events-per-second-on-azure-functions. Accessed 10 April 2019
47. Microsoft, Azure HDInsight documentation (2019), https://docs.microsoft.com/en-us/azure/hdinsight/. Accessed 11 April 2019
48. Microsoft, Apache Spark in Azure HDInsight (2019), https://docs.microsoft.com/en-us/azure/hdinsight/spark/apache-spark-overview. Accessed 10 April 2019
49. Y. Samadi, M. Zbakh, C. Tadonki, Comparative study between hadoop and spark based on hibench benchmarks, in *2016 2nd International Conference on Cloud Computing Technologies and Applications (CloudTech)*, (2016, May), pp. 267–275
50. Microsoft, Real-time streaming in Power BI (2018), https://docs.microsoft.com/en-us/power-bi/service-real-time-streaming. Accessed 11 April 2019
51. Microsoft, Azure time series insights documentation (2017), https://docs.microsoft.com/en-us/azure/time-series-insights. Accessed 11 April 2019
52. Microsoft, Azure Notification Hubs (2019), https://docs.microsoft.com/en-us/azure/notification-hubs. Accessed 11 April 2019
53. B. Saha, K. Goebel, Battery data set, NASA AMES prognostics data repository (2007) https://ti.arc.nasa.gov/tech/dash/groups/pcoe/prognostic-data-repository/
54. Microsoft, Choose the right IoT Hub tier for your solution (2018), https://docs.microsoft.com/en-us/azure/iot-hub/iot-hub-scaling#basic-and-standard-tiers. Accessed 10 May 2019
55. Microsoft, Understand and use Azure IoT Hub SDKs (2019), https://docs.microsoft.com/en-us/azure/iot-hub/iot-hub-devguide-sdks. Accessed 10 May 2019
56. Microsoft, Azure storage account overview (2019), https://docs.microsoft.com/en-us/azure/storage/common/storage-account-overview. Accessed 10 May 2019
57. Microsoft, Reference - IoT Hub endpoints (2019), https://docs.microsoft.com/en-us/azure/iot-hub/iot-hub-devguide-endpoints. Accessed 10 May 2019
58. Microsoft, What is Azure Service Bus? (2018), https://docs.microsoft.com/en-us/azure/service-bus-messaging/service-bus-messaging-overview. Accessed 10 May 2019

59. Microsoft, What is Azure stream analytics (2019), https://docs.microsoft.com/en-us/azure/stream-analytics/stream-analytics-introduction. Accessed 10 May 2019
60. Microsoft, Understand inputs for Azure Stream Analytics (2019), https://docs.microsoft.com/en-us/azure/stream-analytics/stream-analytics-add-inputs. Accessed 12 May 2019
61. Microsoft, Understand outputs from Azure Stream Analytics (2019), https://docs.microsoft.com/en-us/azure/stream-analytics/stream-analytics-define-outputs. Accessed 12 May 2019
62. Microsoft, Azure Functions triggers and bindings concepts (2019), https://docs.microsoft.com/en-us/azure/azure-functions/functions-triggers-bindings. Accessed 12 May 2019
63. Microsoft, Azure Table storage overview (2019), https://docs.microsoft.com/en-us/azure/cosmos-db/table-storage-overview. Accessed 12 May 2019
64. Microsoft, Test a Stream Analytics query with sample data (2018), https://docs.microsoft.com/en-us/azure/stream-analytics/stream-analytics-test-query. Accessed 12 May 2019
65. Microsoft, Supported languages in Azure Functions (2018), https://docs.microsoft.com/en-us/azure/azure-functions/supported-languages. Accessed 12 May 2019
66. Microsoft, Azure Functions C# script (.csx) developer reference (2017), https://docs.microsoft.com/en-us/azure/azure-functions/functions-reference-csharp. Accessed 12 May 2019
67. Microsoft, Azure Table storage bindings for Azure Functions (2018), https://docs.microsoft.com/en-us/azure/azure-functions/functions-bindings-storage-table. Accessed 12 May 2019
68. Microsoft, How to get started with Azure Table storage and Visual Studio connected services (2017), https://docs.microsoft.com/en-us/azure/visual-studio/vs-storage-aspnet5-getting-started-tables. Accessed 12 May 2019
69. S. Athmaja, M. Hanumanthappa, V. Kavitha, A survey of machine learning algo- rithms for big data analytics, in *2017 International Conference on Innovations in Information, Embedded and Communication Systems (ICIIECS)*, (2017, March), pp. 1–4
70. H.U. Dike, Y. Zhou, K.K. Deveerasetty, Q. Wu, Unsupervised learning based on artificial neural network: A review, in *2018 IEEE International Conference on Cyborg and Bionic Systems (CBS)*, (2018, Oct), pp. 322–327
71. D. Silver, T. Hubert, J. Schrittwieser, I. Antonoglou, M. Lai, A. Guez, M. Lanctot, L. Sifre, D. Kumaran, T. Graepel, T.P. Lillicrap, K. Simonyan, D. Hassabis, Mastering chess and shogi by self-play with a general reinforcement learning algorithm. *CoRR*, abs/1712.01815 (2017)
72. Microsoft, What are the Machine Learning products at Microsoft? (2019), https://docs.microsoft.com/en-us/azure/architecture/data-guide/technology-choices/data-science-and-machine-learning?context=azure/machine-   learning/studio/context/ml-context.   Accessed 15 June 2019
73. Microsoft, What is Azure Machine Learning Studio? (2019), https://docs.microsoft.com/en-us/azure/machine-learning/studio/what-is-ml-studio. Accessed 15 June 2019
74. Microsoft, Quickstart: Create your first data science experiment in Azure Machine Learning Studio (2019), https://docs.microsoft.com/en-us/azure/machine-learning/studio/create-experiment. Accessed 15 June 2019
75. Microsoft, Machine Learning module descriptions (2019), https://docs.microsoft.com/en-us/azure/machine-learning/studio-module-reference/machine-learning-module-descriptions. Accessed 15 June 2019
76. Microsoft, Import data (2019), https://docs.microsoft.com/en-us/azure/machine-learning/studio-module-reference/import-data. Accessed 15 June 2019
77. Anup Bhande. What is underfitting and overfitting in machine learning and how to deal with it (2018), https://medium.com/greyatom/what-is-underfitting-and-overfitting-in-machine-learning-and-how-to-deal-with-it-6803a989c76. Accessed 15 June 2019
78. Microsoft, Cross-validate model (2019), https://docs.microsoft.com/en-us/azure/machine-learning/studio-module-reference/cross-validate-model. Accessed 15 June 2019
79. Microsoft, Tutorial 3: Deploy credit risk model - Azure Machine Learning Studio (2019), https://docs.microsoft.com/en-us/azure/machine-learning/studio/tutorial-part3-credit-risk-deploy. Accessed 15 June 2019

80. Microsoft, Azure Machine Learning Studio Web Services: Deployment and consumption (2017), https://docs.microsoft.com/en-us/azure/machine-learning/studio/deploy-consume-web-service-guide. Accessed 15 June 2019

81. Microsoft, What is Azure Machine Learning Service? (2019), https://docs.microsoft.com/en-us/azure/machine-learning/service/overview-what-is-azure-ml. Accessed 15 June 2019

82. Microsoft, Supplemental terms of use for Microsoft Azure previews (2019), https://azure.microsoft.com/en-us/support/legal/preview-supplemental-terms/. Accessed 14 June 2019

83. Microsoft, What is the visual interface for Azure Machine Learning service? (2019), https://docs.microsoft.com/en-us/azure/machine-learning/service/ui-concept-visual-interface. Accessed 15 June 2019

84. Microsoft, What is automated Machine Learning? (2019), https://docs.microsoft.com/en-us/azure/machine-learning/service/concept-automated-ml. Accessed 15 June 2019

85. Microsoft, What is the Azure Machine Learning SDK for Python? (2019), https://docs.microsoft.com/en-us/python/api/overview/azure/ml/intro?view=azure-ml-py. Accessed 15 June 2019

86. Microsoft, The Jupyter notebook introduction (2015), https://jupyter-notebook.readthedocs.io/en/stable/notebook.html. Accessed 15 June 2019

87. Microsoft, Overview of Azure Notebooks (2019), https://docs.microsoft.com/en-us/azure/notebooks/azure-notebooks-overview. Accessed 15 June 2019

88. Microsoft, What is Power BI? (2019), https://docs.microsoft.com/en-us/power-bi/power-bi-overview. Accessed 16 June 2019

89. Microsoft, Query overview in Power BI desktop (2019), https://docs.microsoft.com/en-us/power-bi/desktop-query-overview. Accessed 16 June 2019

90. Microsoft, Comparing Power BI Desktop and the Power BI service (2018), https://docs.microsoft.com/en-us/power-bi/service-service-vs-desktop. Accessed 16 June 2019

91. Microsoft, Basic concepts for designers in the Power BI service (2019), https://docs.microsoft.com/en-us/power-bi/service-basic-concepts. Accessed 16 June 2019

92. Microsoft, Embedded analytics with Power BI (2019), https://docs.microsoft.com/en-us/power-bi/developer/embedding. Accessed 16 June 2019

Appendix A: Regression Example

Linear Regression Matlab Code

```
    function [trainedModel, validationRMSE] = trainRegressionModel
(trainingData)
 inputTable = trainingData;
 predictorNames = {'FF', 'GTCIP', 'GTCIT', 'GTCOP', 'GTCOT', 'GTEGP',
'GTRR', 'GTST', 'HPTP', 'HPTT', 'LP', 'PPT', 'SPT', 'SS', 'TD', 'TIC'};
 predictors = inputTable(:, predictorNames);
 response = inputTable.CD;
 isCategoricalPredictor = [false, false, false, false, false, false,
false, false, false, false, false, false, false, false, false, false];

 % Train a regression model
 % This code specifies all the model options and trains the model.
 concatenatedPredictorsAndResponse = predictors;
 concatenatedPredictorsAndResponse.CD = response;
 linearModel = fitlm(...
    concatenatedPredictorsAndResponse, ...
    'linear', ...
    'RobustOpts', 'off');

 % Create the result struct with predict function
 predictorExtractionFcn = @(t) t(:, predictorNames);
 linearModelPredictFcn = @(x) predict(linearModel, x);
 trainedModel.predictFcn = @(x) linearModelPredictFcn
(predictorExtractionFcn(x));

 % Add additional fields to the result struct
 trainedModel.RequiredVariables = {'FF', 'GTCIP', 'GTCIT', 'GTCOP',
'GTCOT', 'GTEGP', 'GTRR', 'GTST', 'HPTP', 'HPTT', 'LP', 'PPT', 'SPT',
'SS', 'TD', 'TIC'};
 trainedModel.LinearModel = linearModel;
 trainedModel.About = 'This struct is a trained model exported from
Regression Learner R2018a.';
```

© Springer Nature Switzerland AG 2020
F. Balali et al., *Data Intensive Industrial Asset Management*,
https://doi.org/10.1007/978-3-030-35930-0

```
trainedModel.HowToPredict = sprintf('To make predictions on a new
table, T, use: \n yfit = c.predictFcn(T) \nreplacing ''c'' with the name
of the variable that is this struct, e.g. ''trainedModel''. \n \nThe
table, T, must contain the variables returned by: \n c.
RequiredVariables \nVariable formats (e.g. matrix/vector, datatype)
must match the original training data. \nAdditional variables are
ignored. \n \nFor more information, see <a href="matlab:helpview
(fullfile(docroot, ''stats'', ''stats.map''),
''appregression_exportmodeltoworkspace'')">How to predict using an
exported model</a>.');

% Extract predictors and response
% This code processes the data into the right shape for training the
% model.
inputTable = trainingData;
predictorNames = {'FF', 'GTCIP', 'GTCIT', 'GTCOP', 'GTCOT', 'GTEGP',
'GTRR', 'GTST', 'HPTP', 'HPTT', 'LP', 'PPT', 'SPT', 'SS', 'TD', 'TIC'};
predictors = inputTable(:, predictorNames);
response = inputTable.CD;
isCategoricalPredictor = [false, false, false, false, false, false,
false, false, false, false, false, false, false, false, false, false];

% Perform cross-validation
KFolds = 10;
cvp = cvpartition(size(response, 1), 'KFold', KFolds);
% Initialize the predictions to the proper sizes
validationPredictions = response;
for fold = 1:KFolds
  trainingPredictors = predictors(cvp.training(fold), :);
  trainingResponse = response(cvp.training(fold), :);
  foldIsCategoricalPredictor = isCategoricalPredictor;

  % Train a regression model
  % This code specifies all the model options and trains the model.
  concatenatedPredictorsAndResponse = trainingPredictors;
  concatenatedPredictorsAndResponse.CD = trainingResponse;
  linearModel = fitlm(...
    concatenatedPredictorsAndResponse, ...
    'linear', ...
    'RobustOpts', 'off');

  % Create the result struct with predict function
  linearModelPredictFcn = @(x) predict(linearModel, x);
  validationPredictFcn = @(x) linearModelPredictFcn(x);

  % Add additional fields to the result struct

  % Compute validation predictions
  validationPredictors = predictors(cvp.test(fold), :);
  foldPredictions = validationPredictFcn(validationPredictors);
```

```
  % Store predictions in the original order
  validationPredictions(cvp.test(fold), :) = foldPredictions;
end

% Compute validation RMSE
isNotMissing = ~isnan(validationPredictions) & ~isnan(response);
validationRMSE = sqrt(nansum(( validationPredictions - response ).
^2) / numel(response(isNotMissing) ));
```

Linear Interaction Regression Model

```
    function [trainedModel, validationRMSE] = trainRegressionModel
(trainingData)

  % Extract predictors and response
  % This code processes the data into the right shape for training the
  % model.
  inputTable = trainingData;
  predictorNames = {'FF', 'GTCIP', 'GTCIT', 'GTCOP', 'GTCOT', 'GTEGP',
'GTRR', 'GTST', 'HPTP', 'HPTT', 'LP', 'PPT', 'SPT', 'SS', 'TD', 'TIC'};
  predictors = inputTable(:, predictorNames);
  response = inputTable.CD;
  isCategoricalPredictor = [false, false, false, false, false, false,
false, false, false, false, false, false, false, false, false, false];

  % Train a regression model
  % This code specifies all the model options and trains the model.
  concatenatedPredictorsAndResponse = predictors;
  concatenatedPredictorsAndResponse.CD = response;
  linearModel = fitlm(...
    concatenatedPredictorsAndResponse, ...
    'interactions', ...
    'RobustOpts', 'off');

  % Create the result struct with predict function
  predictorExtractionFcn = @(t) t(:, predictorNames);
  linearModelPredictFcn = @(x) predict(linearModel, x);
  trainedModel.predictFcn = @(x) linearModelPredictFcn
(predictorExtractionFcn(x));

  % Add additional fields to the result struct
  trainedModel.RequiredVariables = {'FF', 'GTCIP', 'GTCIT', 'GTCOP',
'GTCOT', 'GTEGP', 'GTRR', 'GTST', 'HPTP', 'HPTT', 'LP', 'PPT', 'SPT',
'SS', 'TD', 'TIC'};
  trainedModel.LinearModel = linearModel;
  trainedModel.About = 'This struct is a trained model exported from
Regression Learner R2018a.';
```

```matlab
trainedModel.HowToPredict = sprintf('To make predictions on a new
table, T, use: \n yfit = c.predictFcn(T) \nreplacing ''c'' with the name
of the variable that is this struct, e.g. ''trainedModel''. \n \nThe
table, T, must contain the variables returned by: \n c.
RequiredVariables \nVariable formats (e.g. matrix/vector, datatype)
must match the original training data. \nAdditional variables are
ignored. \n \nFor more information, see <a href="matlab:helpview
(fullfile(docroot, ''stats'', ''stats.map''),
''appregression_exportmodeltoworkspace'')">How to predict using an
exported model</a>.');


% Extract predictors and response
% This code processes the data into the right shape for training the
% model.
inputTable = trainingData;
predictorNames = {'FF', 'GTCIP', 'GTCIT', 'GTCOP', 'GTCOT', 'GTEGP',
'GTRR', 'GTST', 'HPTP', 'HPTT', 'LP', 'PPT', 'SPT', 'SS', 'TD', 'TIC'};
predictors = inputTable(:, predictorNames);
response = inputTable.CD;
isCategoricalPredictor = [false, false, false, false, false, false,
false, false, false, false, false, false, false, false, false, false];


% Perform cross-validation
KFolds = 10;
cvp = cvpartition(size(response, 1), 'KFold', KFolds);
% Initialize the predictions to the proper sizes
validationPredictions = response;
for fold = 1:KFolds
  trainingPredictors = predictors(cvp.training(fold), :);
  trainingResponse = response(cvp.training(fold), :);
  foldIsCategoricalPredictor = isCategoricalPredictor;


  % Train a regression model
  % This code specifies all the model options and trains the model.
  concatenatedPredictorsAndResponse = trainingPredictors;
  concatenatedPredictorsAndResponse.CD = trainingResponse;
  linearModel = fitlm(...
    concatenatedPredictorsAndResponse, ...
    'interactions', ...
    'RobustOpts', 'off');


  % Create the result struct with predict function
  linearModelPredictFcn = @(x) predict(linearModel, x);
  validationPredictFcn = @(x) linearModelPredictFcn(x);


  % Add additional fields to the result struct


  % Compute validation predictions
  validationPredictors = predictors(cvp.test(fold), :);
  foldPredictions = validationPredictFcn(validationPredictors);
```

```
  % Store predictions in the original order
    validationPredictions(cvp.test(fold), :) = foldPredictions;
  end

  % Compute validation RMSE
  isNotMissing = ~isnan(validationPredictions) & ~isnan(response);
  validationRMSE = sqrt(nansum(( validationPredictions - response ).
^2) / numel(response(isNotMissing) ));
```

Linear Stepwise Regression Model

```
      function [trainedModel, validationRMSE] = trainRegressionModel
(trainingData)

  % Extract predictors and response
  % This code processes the data into the right shape for training the
  % model.
  inputTable = trainingData;
    predictorNames = {'FF', 'GTCIP', 'GTCIT', 'GTCOP', 'GTCOT', 'GTEGP',
'GTRR', 'GTST', 'HPTP', 'HPTT', 'LP', 'PPT', 'SPT', 'SS', 'TD', 'TIC'};
    predictors = inputTable(:, predictorNames);
    response = inputTable.CD;
    isCategoricalPredictor = [false, false, false, false, false, false,
false, false, false, false, false, false, false, false, false, false];

  % Train a regression model
  % This code specifies all the model options and trains the model.
  concatenatedPredictorsAndResponse = predictors;
  concatenatedPredictorsAndResponse.CD = response;
  linearModel = stepwiselm(...
    concatenatedPredictorsAndResponse, ...
    'linear', ...
    'Upper', 'interactions', ...
    'NSteps', 1000, ...
    'Verbose', 0);

  % Create the result struct with predict function
  predictorExtractionFcn = @(t) t(:, predictorNames);
  linearModelPredictFcn = @(x) predict(linearModel, x);
    trainedModel.predictFcn = @(x) linearModelPredictFcn
(predictorExtractionFcn(x));

  % Add additional fields to the result struct
    trainedModel.RequiredVariables = {'FF', 'GTCIP', 'GTCIT', 'GTCOP',
'GTCOT', 'GTEGP', 'GTRR', 'GTST', 'HPTP', 'HPTT', 'LP', 'PPT', 'SPT',
'SS', 'TD', 'TIC'};
    trainedModel.LinearModel = linearModel;
```

```
    trainedModel.About = 'This struct is a trained model exported from
Regression Learner R2018a.';
    trainedModel.HowToPredict = sprintf('To make predictions on a new
table, T, use: \n yfit = c.predictFcn(T) \nreplacing ''c'' with the name
of the variable that is this struct, e.g. ''trainedModel''. \n \nThe
table, T, must contain the variables returned by: \n c.
RequiredVariables \nVariable formats (e.g. matrix/vector, datatype)
must match the original training data. \nAdditional variables are
ignored. \n \nFor more information, see <a href="matlab:helpview
(fullfile(docroot, ''stats'', ''stats.map''),
''appregression_exportmodeltoworkspace'')">How to predict using an
exported model</a>.');


    % Extract predictors and response
    % This code processes the data into the right shape for training the
    % model.
    inputTable = trainingData;
    predictorNames = {'FF', 'GTCIP', 'GTCIT', 'GTCOP', 'GTCOT', 'GTEGP',
'GTRR', 'GTST', 'HPTP', 'HPTT', 'LP', 'PPT', 'SPT', 'SS', 'TD', 'TIC'};
    predictors = inputTable(:, predictorNames);
    response = inputTable.CD;
    isCategoricalPredictor = [false, false, false, false, false, false,
false, false, false, false, false, false, false, false, false, false];


    % Perform cross-validation
    KFolds = 10;
    cvp = cvpartition(size(response, 1), 'KFold', KFolds);
    % Initialize the predictions to the proper sizes
    validationPredictions = response;
    for fold = 1:KFolds
        trainingPredictors = predictors(cvp.training(fold), :);
        trainingResponse = response(cvp.training(fold), :);
        foldIsCategoricalPredictor = isCategoricalPredictor;


        % Train a regression model
        % This code specifies all the model options and trains the model.
        concatenatedPredictorsAndResponse = trainingPredictors;
        concatenatedPredictorsAndResponse.CD = trainingResponse;
        linearModel = stepwiselm(...
            concatenatedPredictorsAndResponse, ...
            'linear', ...
            'Upper', 'interactions', ...
            'NSteps', 1000, ...
            'Verbose', 0);


        % Create the result struct with predict function
        linearModelPredictFcn = @(x) predict(linearModel, x);
        validationPredictFcn = @(x) linearModelPredictFcn(x);


        % Add additional fields to the result struct
```

```
    % Compute validation predictions
    validationPredictors = predictors(cvp.test(fold), :);
    foldPredictions = validationPredictFcn(validationPredictors);

    % Store predictions in the original order
    validationPredictions(cvp.test(fold), :) = foldPredictions;
  end

  % Compute validation RMSE
  isNotMissing = ~isnan(validationPredictions) & ~isnan(response);
  validationRMSE = sqrt(nansum(( validationPredictions - response ).
^2) / numel(response(isNotMissing) ));
```

Decision Tress

```
    function [trainedModel, validationRMSE] = trainRegressionModel
(trainingData)

  % Extract predictors and response
  % This code processes the data into the right shape for training the
  % model.
  inputTable = trainingData;
  predictorNames = {'FF', 'GTCIP', 'GTCIT', 'GTCOP', 'GTCOT', 'GTEGP',
'GTRR', 'GTST', 'HPTP', 'HPTT', 'LP', 'PPT', 'SPT', 'SS', 'TD', 'TIC'};
  predictors = inputTable(:, predictorNames);
  response = inputTable.CD;
  isCategoricalPredictor = [false, false, false, false, false, false,
false, false, false, false, false, false, false, false, false, false];

  % Train a regression model
  % This code specifies all the model options and trains the model.
  regressionTree = fitrtree(...
    predictors, ...
    response, ...
    'MinLeafSize', 4, ...
    'Surrogate', 'off');

  % Create the result struct with predict function
  predictorExtractionFcn = @(t) t(:, predictorNames);
  treePredictFcn = @(x) predict(regressionTree, x);
  trainedModel.predictFcn = @(x) treePredictFcn
(predictorExtractionFcn(x));

  % Add additional fields to the result struct
  trainedModel.RequiredVariables = {'FF', 'GTCIP', 'GTCIT', 'GTCOP',
'GTCOT', 'GTEGP', 'GTRR', 'GTST', 'HPTP', 'HPTT', 'LP', 'PPT', 'SPT',
'SS', 'TD', 'TIC'};
```

```
trainedModel.RegressionTree = regressionTree;
trainedModel.About = 'This struct is a trained model exported from
Regression Learner R2018a.';
trainedModel.HowToPredict = sprintf('To make predictions on a new
table, T, use: \n yfit = c.predictFcn(T) \nreplacing ''c'' with the name
of the variable that is this struct, e.g. ''trainedModel''. \n \nThe
table, T, must contain the variables returned by: \n c.
RequiredVariables \nVariable formats (e.g. matrix/vector, datatype)
must match the original training data. \nAdditional variables are
ignored. \n \nFor more information, see <a href="matlab:helpview
(fullfile(docroot, ''stats'', ''stats.map''),
''appregression_exportmodeltoworkspace'')">How to predict using an
exported model</a>.');

% Extract predictors and response
% This code processes the data into the right shape for training the
% model.
inputTable = trainingData;
predictorNames = {'FF', 'GTCIP', 'GTCIT', 'GTCOP', 'GTCOT', 'GTEGP',
'GTRR', 'GTST', 'HPTP', 'HPTT', 'LP', 'PPT', 'SPT', 'SS', 'TD', 'TIC'};
predictors = inputTable(:, predictorNames);
response = inputTable.CD;
isCategoricalPredictor = [false, false, false, false, false, false,
false, false, false, false, false, false, false, false, false, false];

% Perform cross-validation
partitionedModel = crossval(trainedModel.RegressionTree, 'KFold',
10);

% Compute validation predictions
validationPredictions = kfoldPredict(partitionedModel);

% Compute validation RMSE
validationRMSE = sqrt(kfoldLoss(partitionedModel, 'LossFun',
'mse'));
```

Bagged Decision Tree

```
    function [trainedModel, validationRMSE] = trainRegressionModel
(trainingData)

% Extract predictors and response
% This code processes the data into the right shape for training the
% model.
inputTable = trainingData;
predictorNames = {'FF', 'GTCIP', 'GTCIT', 'GTCOP', 'GTCOT', 'GTEGP',
'GTRR', 'GTST', 'HPTP', 'HPTT', 'LP', 'PPT', 'SPT', 'SS', 'TD', 'TIC'};
```

```
predictors = inputTable(:, predictorNames);
response = inputTable.CD;
isCategoricalPredictor = [false, false, false, false, false, false,
false, false, false, false, false, false, false, false, false, false];

% Train a regression model
% This code specifies all the model options and trains the model.
template = templateTree(...
  'MinLeafSize', 8);
regressionEnsemble = fitrensemble(...
  predictors, ...
  response, ...
  'Method', 'Bag', ...
  'NumLearningCycles', 30, ...
  'Learners', template);

% Create the result struct with predict function
predictorExtractionFcn = @(t) t(:, predictorNames);
ensemblePredictFcn = @(x) predict(regressionEnsemble, x);
trainedModel.predictFcn = @(x) ensemblePredictFcn
(predictorExtractionFcn(x));

% Add additional fields to the result struct
trainedModel.RequiredVariables = {'FF', 'GTCIP', 'GTCIT', 'GTCOP',
'GTCOT', 'GTEGP', 'GTRR', 'GTST', 'HPTP', 'HPTT', 'LP', 'PPT', 'SPT',
'SS', 'TD', 'TIC'};
trainedModel.RegressionEnsemble = regressionEnsemble;
trainedModel.About = 'This struct is a trained model exported from
Regression Learner R2018a.';
trainedModel.HowToPredict = sprintf('To make predictions on a new
table, T, use: \n yfit = c.predictFcn(T) \nreplacing ''c'' with the name
of the variable that is this struct, e.g. ''trainedModel''. \n \nThe
table, T, must contain the variables returned by: \n c.
RequiredVariables \nVariable formats (e.g. matrix/vector, datatype)
must match the original training data. \nAdditional variables are
ignored. \n \nFor more information, see <a href="matlab:helpview
(fullfile(docroot, ''stats'', ''stats.map''),
''appregression_exportmodeltoworkspace'')">How to predict using an
exported model</a>.');

% Extract predictors and response
% This code processes the data into the right shape for training the
% model.
inputTable = trainingData;
predictorNames = {'FF', 'GTCIP', 'GTCIT', 'GTCOP', 'GTCOT', 'GTEGP',
'GTRR', 'GTST', 'HPTP', 'HPTT', 'LP', 'PPT', 'SPT', 'SS', 'TD', 'TIC'};
predictors = inputTable(:, predictorNames);
response = inputTable.CD;
isCategoricalPredictor = [false, false, false, false, false, false,
false, false, false, false, false, false, false, false, false, false];
```

```
% Perform cross-validation
partitionedModel = crossval(trainedModel.RegressionEnsemble,
'KFold', 10);

% Compute validation predictions
validationPredictions = kfoldPredict(partitionedModel);

% Compute validation RMSE
validationRMSE = sqrt(kfoldLoss(partitionedModel, 'LossFun',
'mse'));
```

Linear SVM

```
    function [trainedModel, validationRMSE] = trainRegressionModel
(trainingData)

  % Extract predictors and response
  % This code processes the data into the right shape for training the
  % model.
  inputTable = trainingData;
  predictorNames = {'FF', 'GTCIP', 'GTCIT', 'GTCOP', 'GTCOT', 'GTEGP',
'GTRR', 'GTST', 'HPTP', 'HPTT', 'LP', 'PPT', 'SPT', 'SS', 'TD', 'TIC'};
  predictors = inputTable(:, predictorNames);
  response = inputTable.CD;
  isCategoricalPredictor = [false, false, false, false, false, false,
false, false, false, false, false, false, false, false, false, false];

  % Train a regression model
  % This code specifies all the model options and trains the model.
  responseScale = iqr(response);
  if ~isfinite(responseScale) || responseScale == 0.0
    responseScale = 1.0;
  end
  boxConstraint = responseScale/1.349;
  epsilon = responseScale/13.49;
  regressionSVM = fitrsvm(...
    predictors, ...
    response, ...
    'KernelFunction', 'linear', ...
    'PolynomialOrder', [], ...
    'KernelScale', 'auto', ...
    'BoxConstraint', boxConstraint, ...
    'Epsilon', epsilon, ...
    'Standardize', true);

  % Create the result struct with predict function
  predictorExtractionFcn = @(t) t(:, predictorNames);
```

```
  svmPredictFcn = @(x) predict(regressionSVM, x);
  trainedModel.predictFcn = @(x) svmPredictFcn
(predictorExtractionFcn(x));


% Add additional fields to the result struct
  trainedModel.RequiredVariables = {'FF', 'GTCIP', 'GTCIT', 'GTCOP',
'GTCOT', 'GTEGP', 'GTRR', 'GTST', 'HPTP', 'HPTT', 'LP', 'PPT', 'SPT',
'SS', 'TD', 'TIC'};
  trainedModel.RegressionSVM = regressionSVM;
  trainedModel.About = 'This struct is a trained model exported from
Regression Learner R2018a.';
  trainedModel.HowToPredict = sprintf('To make predictions on a new
table, T, use: \n yfit = c.predictFcn(T) \nreplacing ''c'' with the name
of the variable that is this struct, e.g. ''trainedModel''. \n \nThe
table, T, must contain the variables returned by: \n c.
RequiredVariables \nVariable formats (e.g. matrix/vector, datatype)
must match the original training data. \nAdditional variables are
ignored. \n \nFor more information, see <a href="matlab:helpview
(fullfile(docroot, ''stats'', ''stats.map''),
''appregression_exportmodeltoworkspace'')">How to predict using an
exported model</a>.');


% Extract predictors and response
% This code processes the data into the right shape for training the
% model.
  inputTable = trainingData;
  predictorNames = {'FF', 'GTCIP', 'GTCIT', 'GTCOP', 'GTCOT', 'GTEGP',
'GTRR', 'GTST', 'HPTP', 'HPTT', 'LP', 'PPT', 'SPT', 'SS', 'TD', 'TIC'};
  predictors = inputTable(:, predictorNames);
  response = inputTable.CD;
  isCategoricalPredictor = [false, false, false, false, false, false,
false, false, false, false, false, false, false, false, false, false];


% Perform cross-validation
  KFolds = 10;
  cvp = cvpartition(size(response, 1), 'KFold', KFolds);
% Initialize the predictions to the proper sizes
  validationPredictions = response;
  for fold = 1:KFolds
    trainingPredictors = predictors(cvp.training(fold), :);
    trainingResponse = response(cvp.training(fold), :);
    foldIsCategoricalPredictor = isCategoricalPredictor;


  % Train a regression model
  % This code specifies all the model options and trains the model.
  responseScale = iqr(trainingResponse);
  if ~isfinite(responseScale) || responseScale == 0.0
    responseScale = 1.0;
  end
  boxConstraint = responseScale/1.349;
  epsilon = responseScale/13.49;
  regressionSVM = fitrsvm(...
    trainingPredictors, ...
```

```
        trainingResponse, ...
        'KernelFunction', 'linear', ...
        'PolynomialOrder', [], ...
        'KernelScale', 'auto', ...
        'BoxConstraint', boxConstraint, ...
        'Epsilon', epsilon, ...
        'Standardize', true);

    % Create the result struct with predict function
    svmPredictFcn = @(x) predict(regressionSVM, x);
    validationPredictFcn = @(x) svmPredictFcn(x);

    % Add additional fields to the result struct

    % Compute validation predictions
    validationPredictors = predictors(cvp.test(fold), :);
    foldPredictions = validationPredictFcn(validationPredictors);

    % Store predictions in the original order
    validationPredictions(cvp.test(fold), :) = foldPredictions;
  end

  % Compute validation RMSE
  isNotMissing = ~isnan(validationPredictions) & ~isnan(response);
  validationRMSE = sqrt(nansum(( validationPredictions - response ).
^2) / numel(response(isNotMissing) ));
```

Appendix B: Classification Example

Fine Tree

```
function [trainedClassifier, validationAccuracy] = trainClassifier
(trainingData)
inputTable = trainingData;
predictorNames = {'tau1', 'tau2', 'tau3', 'tau4', 'p1', 'p2', 'p3',
'p4', 'g1', 'g2', 'g3', 'g4'};
predictors = inputTable(:, predictorNames);
response = inputTable.stabf;
isCategoricalPredictor = [false, false, false, false, false, false,
false, false, false, false, false, false];

% Train a classifier
% This code specifies all the classifier options and trains the classifier.
classificationTree = fitctree(...
   predictors, ...
   response, ...
   'SplitCriterion', 'gdi', ...
   'MaxNumSplits', 100, ...
   'Surrogate', 'off', ...
   'ClassNames', categorical({'stable'; 'unstable'}));

% Create the result struct with predict function
predictorExtractionFcn = @(t) t(:, predictorNames);
treePredictFcn = @(x) predict(classificationTree, x);
trainedClassifier.predictFcn = @(x) treePredictFcn
(predictorExtractionFcn(x));

% Add additional fields to the result struct
trainedClassifier.RequiredVariables = {'tau1', 'tau2', 'tau3', 'tau4',
'p1', 'p2', 'p3', 'p4', 'g1', 'g2', 'g3', 'g4'};
trainedClassifier.ClassificationTree = classificationTree;
trainedClassifier.About = 'This struct is a trained model exported from
Classification Learner R2018a.';
```

© Springer Nature Switzerland AG 2020
F. Balali et al., *Data Intensive Industrial Asset Management*,
https://doi.org/10.1007/978-3-030-35930-0

```
trainedClassifier.HowToPredict = sprintf('To make predictions on a new
table, T, use: \n yfit = c.predictFcn(T) \nreplacing ''c'' with the name
of the variable that is this struct, e.g. ''trainedModel''. \n \nThe
table, T, must contain the variables returned by: \n c.
RequiredVariables \nVariable formats (e.g. matrix/vector, datatype)
must match the original training data. \nAdditional variables are
ignored. \n \nFor more information, see <a href="matlab:helpview
(fullfile(docroot, ''stats'', ''stats.map''),
''appclassification_exportmodeltoworkspace'')">How to predict using
an exported model</a>.');

% Extract predictors and response
% This code processes the data into the right shape for training the
% model.
inputTable = trainingData;
predictorNames = {'tau1', 'tau2', 'tau3', 'tau4', 'p1', 'p2', 'p3',
'p4', 'g1', 'g2', 'g3', 'g4'};
predictors = inputTable(:, predictorNames);
response = inputTable.stabf;
isCategoricalPredictor = [false, false, false, false, false, false,
false, false, false, false, false, false];

% Perform cross-validation
partitionedModel = crossval(trainedClassifier.ClassificationTree,
'KFold', 10);

% Compute validation predictions
[validationPredictions, validationScores] = kfoldPredict
(partitionedModel);

% Compute validation accuracy
validationAccuracy = 1 - kfoldLoss(partitionedModel, 'LossFun',
'ClassifError');
```

Bagged Tree

```
function [trainedClassifier, validationAccuracy] = trainClassifier
(trainingData)
inputTable = trainingData;
predictorNames = {'tau1', 'tau2', 'tau3', 'tau4', 'p1', 'p2', 'p3',
'p4', 'g1', 'g2', 'g3', 'g4'};
predictors = inputTable(:, predictorNames);
response = inputTable.stabf;
isCategoricalPredictor = [false, false, false, false, false, false,
false, false, false, false, false, false];
```

```
% Train a classifier
% This code specifies all the classifier options and trains the classifier.
template = templateTree(...
  'MaxNumSplits', 9999);
classificationEnsemble = fitcensemble(...
  predictors, ...
  response, ...
  'Method', 'Bag', ...
  'NumLearningCycles', 30, ...
  'Learners', template, ...
  'ClassNames', categorical({'stable'; 'unstable'}));

% Create the result struct with predict function
predictorExtractionFcn = @(t) t(:, predictorNames);
ensemblePredictFcn = @(x) predict(classificationEnsemble, x);
trainedClassifier.predictFcn = @(x) ensemblePredictFcn
(predictorExtractionFcn(x));

% Add additional fields to the result struct
trainedClassifier.RequiredVariables = {'tau1', 'tau2', 'tau3', 'tau4',
'p1', 'p2', 'p3', 'p4', 'g1', 'g2', 'g3', 'g4'};
trainedClassifier.ClassificationEnsemble = classificationEnsemble;
trainedClassifier.About = 'This struct is a trained model exported from
Classification Learner R2018a.';
trainedClassifier.HowToPredict = sprintf('To make predictions on a new
table, T, use: \n yfit = c.predictFcn(T) \nreplacing ''c'' with the name
of the variable that is this struct, e.g. ''trainedModel''. \n \nThe
table, T, must contain the variables returned by: \n c.
RequiredVariables \nVariable formats (e.g. matrix/vector, datatype)
must match the original training data. \nAdditional variables are
ignored. \n \nFor more information, see <a href="matlab:helpview
(fullfile(docroot, ''stats'', ''stats.map''),
''appclassification_exportmodeltoworkspace'')">How to predict using
an exported model</a>.');

% Extract predictors and response
% This code processes the data into the right shape for training the
% model.
inputTable = trainingData;
predictorNames = {'tau1', 'tau2', 'tau3', 'tau4', 'p1', 'p2', 'p3',
'p4', 'g1', 'g2', 'g3', 'g4'};
predictors = inputTable(:, predictorNames);
response = inputTable.stabf;
isCategoricalPredictor = [false, false, false, false, false, false,
false, false, false, false, false, false];

% Perform cross-validation
partitionedModel = crossval(trainedClassifier.ClassificationEnsemble,
'KFold', 10);
```

```
% Compute validation predictions
[validationPredictions, validationScores] = kfoldPredict
(partitionedModel);

% Compute validation accuracy
validationAccuracy = 1 - kfoldLoss(partitionedModel, 'LossFun',
'ClassifError');
```

Fine KNN

```
function [trainedClassifier, validationAccuracy] = trainClassifier
(trainingData)
inputTable = trainingData;
predictorNames = {'tau1', 'tau2', 'tau3', 'tau4', 'p1', 'p2', 'p3',
'p4', 'g1', 'g2', 'g3', 'g4'};
predictors = inputTable(:, predictorNames);
response = inputTable.stabf;
isCategoricalPredictor = [false, false, false, false, false, false,
false, false, false, false, false, false];

% Train a classifier
% This code specifies all the classifier options and trains the classifier.
classificationKNN = fitcknn(...
  predictors, ...
  response, ...
  'Distance', 'Euclidean', ...
  'Exponent', [], ...
  'NumNeighbors', 1, ...
  'DistanceWeight', 'Equal', ...
  'Standardize', true, ...
  'ClassNames', categorical({'stable'; 'unstable'}));

% Create the result struct with predict function
predictorExtractionFcn = @(t) t(:, predictorNames);
knnPredictFcn = @(x) predict(classificationKNN, x);
trainedClassifier.predictFcn = @(x) knnPredictFcn
(predictorExtractionFcn(x));

% Add additional fields to the result struct
trainedClassifier.RequiredVariables = {'tau1', 'tau2', 'tau3', 'tau4',
'p1', 'p2', 'p3', 'p4', 'g1', 'g2', 'g3', 'g4'};
trainedClassifier.ClassificationKNN = classificationKNN;
trainedClassifier.About = 'This struct is a trained model exported from
Classification Learner R2018a.';
trainedClassifier.HowToPredict = sprintf('To make predictions on a new
table, T, use: \n yfit = c.predictFcn(T) \nreplacing ''c'' with the name
of the variable that is this struct, e.g. ''trainedModel''. \n \nThe
```

```
table, T, must contain the variables returned by: \n c.
RequiredVariables \nVariable formats (e.g. matrix/vector, datatype)
must match the original training data. \nAdditional variables are
ignored. \n \nFor more information, see <a href="matlab:helpview
(fullfile(docroot, ''stats''), ''stats.map''),
''appclassification_exportmodeltoworkspace'')">How to predict using
an exported model</a>.');

% Extract predictors and response
% This code processes the data into the right shape for training the
% model.
inputTable = trainingData;
predictorNames = {'tau1', 'tau2', 'tau3', 'tau4', 'p1', 'p2', 'p3',
'p4', 'g1', 'g2', 'g3', 'g4'};
predictors = inputTable(:, predictorNames);
response = inputTable.stabf;
isCategoricalPredictor = [false, false, false, false, false, false,
false, false, false, false, false, false];

% Perform cross-validation
partitionedModel = crossval(trainedClassifier.ClassificationKNN,
'KFold', 10);

% Compute validation predictions
[validationPredictions, validationScores] = kfoldPredict
(partitionedModel);

% Compute validation accuracy
validationAccuracy = 1 - kfoldLoss(partitionedModel, 'LossFun',
'ClassifError');
```

Weighted KNN

```
function [trainedClassifier, validationAccuracy] = trainClassifier
(trainingData)
inputTable = trainingData;
predictorNames = {'tau1', 'tau2', 'tau3', 'tau4', 'p1', 'p2', 'p3',
'p4', 'g1', 'g2', 'g3', 'g4'};
predictors = inputTable(:, predictorNames);
response = inputTable.stabf;
isCategoricalPredictor = [false, false, false, false, false, false,
false, false, false, false, false, false];

% Train a classifier
% This code specifies all the classifier options and trains the classifier.
classificationKNN = fitcknn(...
  predictors, ...
```

```
    response, ...
    'Distance', 'Euclidean', ...
    'Exponent', [], ...
    'NumNeighbors', 10, ...
    'DistanceWeight', 'SquaredInverse', ...
    'Standardize', true, ...
    'ClassNames', categorical({'stable'; 'unstable'}));

% Create the result struct with predict function
predictorExtractionFcn = @(t) t(:, predictorNames);
knnPredictFcn = @(x) predict(classificationKNN, x);
trainedClassifier.predictFcn = @(x) knnPredictFcn
(predictorExtractionFcn(x));

% Add additional fields to the result struct
trainedClassifier.RequiredVariables = {'tau1', 'tau2', 'tau3', 'tau4',
'p1', 'p2', 'p3', 'p4', 'g1', 'g2', 'g3', 'g4'};
trainedClassifier.ClassificationKNN = classificationKNN;
trainedClassifier.About = 'This struct is a trained model exported from
Classification Learner R2018a.';
trainedClassifier.HowToPredict = sprintf('To make predictions on a new
table, T, use: \n yfit = c.predictFcn(T) \nreplacing ''c'' with the name
of the variable that is this struct, e.g. ''trainedModel''. \n \nThe
table, T, must contain the variables returned by: \n c.
RequiredVariables \nVariable formats (e.g. matrix/vector, datatype)
must match the original training data. \nAdditional variables are
ignored. \n \nFor more information, see <a href="matlab:helpview
(fullfile(docroot, ''stats'', ''stats.map''),
''appclassification_exportmodeltoworkspace'') ">How to predict using
an exported model</a>.');

% Extract predictors and response
% This code processes the data into the right shape for training the
% model.
inputTable = trainingData;
predictorNames = {'tau1', 'tau2', 'tau3', 'tau4', 'p1', 'p2', 'p3',
'p4', 'g1', 'g2', 'g3', 'g4'};
predictors = inputTable(:, predictorNames);
response = inputTable.stabf;
isCategoricalPredictor = [false, false, false, false, false, false,
false, false, false, false, false, false];

% Perform cross-validation
partitionedModel = crossval(trainedClassifier.ClassificationKNN,
'KFold', 10);

% Compute validation predictions
[validationPredictions, validationScores] = kfoldPredict
(partitionedModel);
```

```
% Compute validation accuracy
validationAccuracy = 1 - kfoldLoss(partitionedModel, 'LossFun',
'ClassifError');
```

Logistic Regression

```
function [trainedClassifier, validationAccuracy] = trainClassifier
(trainingData)
inputTable = trainingData;
predictorNames = {'tau1', 'tau2', 'tau3', 'tau4', 'p1', 'p2', 'p3',
'p4', 'g1', 'g2', 'g3', 'g4'};
predictors = inputTable(:, predictorNames);
response = inputTable.stabf;
isCategoricalPredictor = [false, false, false, false, false, false,
false, false, false, false, false, false];

% Train a classifier
% This code specifies all the classifier options and trains the classifier.
% For logistic regression, the response values must be converted to zeros
% and ones because the responses are assumed to follow a binomial
% distribution.
% 1 or true = 'successful' class
% 0 or false = 'failure' class
% NaN - missing response.
successClass = 'stable';
failureClass = 'unstable';
% Compute the majority response class. If there is a NaN-prediction from
% fitglm, convert NaN to this majority class label.
numSuccess = sum(response == successClass);
numFailure = sum(response == failureClass);
if numSuccess > numFailure
  missingClass = successClass;
else
  missingClass = failureClass;
end
responseCategories = {successClass, failureClass};
successFailureAndMissingClasses = categorical({successClass;
failureClass; missingClass}, responseCategories);
isMissing = isundefined(response);
zeroOneResponse = double(ismember(response, successClass));
zeroOneResponse(isMissing) = NaN;
% Prepare input arguments to fitglm.
concatenatedPredictorsAndResponse = [predictors, table
(zeroOneResponse)];
% Train using fitglm.
GeneralizedLinearModel = fitglm(...
  concatenatedPredictorsAndResponse, ...
```

```
'Distribution', 'binomial', ...
'link', 'logit');


% Convert predicted probabilities to predicted class labels and scores.
convertSuccessProbsToPredictions = @
(p) successFailureAndMissingClasses( ~isnan(p).*( (p<0.5) +1 ) + isnan
(p)*3 );
returnMultipleValuesFcn = @(varargin) varargin{1:max(1,nargout)};
scoresFcn = @(p) [p, 1-p];
predictionsAndScoresFcn = @(p) returnMultipleValuesFcn(
convertSuccessProbsToPredictions(p), scoresFcn(p) );


% Create the result struct with predict function
predictorExtractionFcn = @(t) t(:, predictorNames);
logisticRegressionPredictFcn = @(x) predictionsAndScoresFcn( predict
(GeneralizedLinearModel, x) );
trainedClassifier.predictFcn = @(x) logisticRegressionPredictFcn
(predictorExtractionFcn(x));


% Add additional fields to the result struct
trainedClassifier.RequiredVariables = {'tau1', 'tau2', 'tau3', 'tau4',
'p1', 'p2', 'p3', 'p4', 'g1', 'g2', 'g3', 'g4'};
trainedClassifier.GeneralizedLinearModel = GeneralizedLinearModel;
trainedClassifier.SuccessClass = successClass;
trainedClassifier.FailureClass = failureClass;
trainedClassifier.MissingClass = missingClass;
trainedClassifier.ClassNames = {successClass; failureClass};
trainedClassifier.About = 'This struct is a trained model exported from
Classification Learner R2018a.';
trainedClassifier.HowToPredict = sprintf('To make predictions on a new
table, T, use: \n yfit = c.predictFcn(T) \nreplacing ''c'' with the name
of the variable that is this struct, e.g. ''trainedModel''. \n \nThe
table, T, must contain the variables returned by: \n c.
RequiredVariables \nVariable formats (e.g. matrix/vector, datatype)
must match the original training data. \nAdditional variables are
ignored. \n \nFor more information, see <a href="matlab:helpview
(fullfile(docroot, ''stats'', ''stats.map''),
''appclassification_exportmodeltoworkspace'')">How to predict using
an exported model</a>.');


% Extract predictors and response
% This code processes the data into the right shape for training the
% model.
inputTable = trainingData;
predictorNames = {'tau1', 'tau2', 'tau3', 'tau4', 'p1', 'p2', 'p3',
'p4', 'g1', 'g2', 'g3', 'g4'};
predictors = inputTable(:, predictorNames);
response = inputTable.stabf;
isCategoricalPredictor = [false, false, false, false, false, false,
false, false, false, false, false, false];
```

```
% Perform cross-validation
KFolds = 10;
cvp = cvpartition(response, 'KFold', KFolds);
% Initialize the predictions to the proper sizes
validationPredictions = response;
numObservations = size(predictors, 1);
numClasses = 2;
validationScores = NaN(numObservations, numClasses);
for fold = 1:KFolds
  trainingPredictors = predictors(cvp.training(fold), :);
  trainingResponse = response(cvp.training(fold), :);
  foldIsCategoricalPredictor = isCategoricalPredictor;

  % Train a classifier
  % This code specifies all the classifier options and trains the
classifier.
  % For logistic regression, the response values must be converted to
zeros
  % and ones because the responses are assumed to follow a binomial
  % distribution.
  % 1 or true = 'successful' class
  % 0 or false = 'failure' class
  % NaN - missing response.
  successClass = 'stable';
  failureClass = 'unstable';
  % Compute the majority response class. If there is a NaN-prediction
from
  % fitglm, convert NaN to this majority class label.
  numSuccess = sum(trainingResponse == successClass);
  numFailure = sum(trainingResponse == failureClass);
  if numSuccess > numFailure
    missingClass = successClass;
  else
    missingClass = failureClass;
  end
  responseCategories = {successClass, failureClass};
  successFailureAndMissingClasses = categorical({successClass;
failureClass; missingClass}, responseCategories);
  isMissing = isundefined(trainingResponse);
  zeroOneResponse = double(ismember(trainingResponse,
successClass));
  zeroOneResponse(isMissing) = NaN;
  % Prepare input arguments to fitglm.
  concatenatedPredictorsAndResponse = [trainingPredictors, table
(zeroOneResponse)];
  % Train using fitglm.
  GeneralizedLinearModel = fitglm(...
    concatenatedPredictorsAndResponse, ...
    'Distribution', 'binomial', ...
    'link', 'logit');

  % Convert predicted probabilities to predicted class labels and
scores.
```

```
  convertSuccessProbsToPredictions = @
(p) successFailureAndMissingClasses ( ~isnan(p) .* ( (p<0.5) + 1 ) + isnan
(p)*3 ) ;
  returnMultipleValuesFcn = @(varargin) varargin{1:max(1,nargout)};
  scoresFcn = @(p) [p, 1-p] ;
  predictionsAndScoresFcn = @(p) returnMultipleValuesFcn(
convertSuccessProbsToPredictions(p), scoresFcn(p) ) ;

  % Create the result struct with predict function
  logisticRegressionPredictFcn = @(x) predictionsAndScoresFcn(
predict(GeneralizedLinearModel, x) ) ;
  validationPredictFcn = @(x) logisticRegressionPredictFcn(x) ;

  % Add additional fields to the result struct

  % Compute validation predictions
  validationPredictors = predictors(cvp.test(fold), :) ;
  [foldPredictions, foldScores] = validationPredictFcn
(validationPredictors) ;

  % Store predictions in the original order
  validationPredictions(cvp.test(fold), :) = foldPredictions;
  validationScores(cvp.test(fold), :) = foldScores;
end

% Compute validation accuracy
correctPredictions = (validationPredictions == response) ;
isMissing = ismissing(response) ;
correctPredictions = correctPredictions(~isMissing) ;
validationAccuracy = sum(correctPredictions)/length
(correctPredictions) ;
```

Linear SVM

```
function [trainedClassifier, validationAccuracy] = trainClassifier
(trainingData)
inputTable = trainingData;
predictorNames = {'tau1', 'tau2', 'tau3', 'tau4', 'p1', 'p2', 'p3',
'p4', 'g1', 'g2', 'g3', 'g4'};
predictors = inputTable(:, predictorNames) ;
response = inputTable.stabf;
isCategoricalPredictor = [false, false, false, false, false, false,
false, false, false, false, false, false] ;

% Train a classifier
% This code specifies all the classifier options and trains the classifier.
classificationSVM = fitcsvm( ...
```

```
predictors, ...
response, ...
'KernelFunction', 'linear', ...
'PolynomialOrder', [], ...
'KernelScale', 'auto', ...
'BoxConstraint', 1, ...
'Standardize', true, ...
'ClassNames', categorical({'stable'; 'unstable'}));


% Create the result struct with predict function
predictorExtractionFcn = @(t) t(:, predictorNames);
svmPredictFcn = @(x) predict(classificationSVM, x);
trainedClassifier.predictFcn = @(x) svmPredictFcn
(predictorExtractionFcn(x));


% Add additional fields to the result struct
trainedClassifier.RequiredVariables = {'tau1', 'tau2', 'tau3', 'tau4',
'p1', 'p2', 'p3', 'p4', 'g1', 'g2', 'g3', 'g4'};
trainedClassifier.ClassificationSVM = classificationSVM;
trainedClassifier.About = 'This struct is a trained model exported from
Classification Learner R2018a.';
trainedClassifier.HowToPredict = sprintf('To make predictions on a new
table, T, use: \n yfit = c.predictFcn(T) \nreplacing ''c'' with the name
of the variable that is this struct, e.g. ''trainedModel''. \n \nThe
table, T, must contain the variables returned by: \n c.
RequiredVariables \nVariable formats (e.g. matrix/vector, datatype)
must match the original training data. \nAdditional variables are
ignored. \n \nFor more information, see <a href="matlab:helpview
(fullfile(docroot, ''stats'', ''stats.map''),
''appclassification_exportmodeltoworkspace'')">How to predict using
an exported model</a>.');


% Extract predictors and response
% This code processes the data into the right shape for training the
% model.
inputTable = trainingData;
predictorNames = {'tau1', 'tau2', 'tau3', 'tau4', 'p1', 'p2', 'p3',
'p4', 'g1', 'g2', 'g3', 'g4'};
predictors = inputTable(:, predictorNames);
response = inputTable.stabf;
isCategoricalPredictor = [false, false, false, false, false, false,
false, false, false, false, false, false];


% Perform cross-validation
partitionedModel = crossval(trainedClassifier.ClassificationSVM,
'KFold', 10);


% Compute validation predictions
[validationPredictions, validationScores] = kfoldPredict
(partitionedModel);
```

```
% Compute validation accuracy
validationAccuracy = 1 - kfoldLoss(partitionedModel, 'LossFun',
'ClassifError');
```

Quadratic SVM

```
function [trainedClassifier, validationAccuracy] = trainClassifier
(trainingData)
inputTable = trainingData;
predictorNames = {'tau1', 'tau2', 'tau3', 'tau4', 'p1', 'p2', 'p3',
'p4', 'g1', 'g2', 'g3', 'g4'};
predictors = inputTable(:, predictorNames);
response = inputTable.stabf;
isCategoricalPredictor = [false, false, false, false, false, false,
false, false, false, false, false, false];

% Train a classifier
% This code specifies all the classifier options and trains the classifier.
classificationSVM = fitcsvm(...
  predictors, ...
  response, ...
  'KernelFunction', 'polynomial', ...
  'PolynomialOrder', 2, ...
  'KernelScale', 'auto', ...
  'BoxConstraint', 1, ...
  'Standardize', true, ...
  'ClassNames', categorical({'stable'; 'unstable'}));

% Create the result struct with predict function
predictorExtractionFcn = @(t) t(:, predictorNames);
svmPredictFcn = @(x) predict(classificationSVM, x);
trainedClassifier.predictFcn = @(x) svmPredictFcn
(predictorExtractionFcn(x));

% Add additional fields to the result struct
trainedClassifier.RequiredVariables = {'tau1', 'tau2', 'tau3', 'tau4',
'p1', 'p2', 'p3', 'p4', 'g1', 'g2', 'g3', 'g4'};
trainedClassifier.ClassificationSVM = classificationSVM;
trainedClassifier.About = 'This struct is a trained model exported from
Classification Learner R2018a.';
trainedClassifier.HowToPredict = sprintf('To make predictions on a new
table, T, use: \n yfit = c.predictFcn(T) \nreplacing ''c'' with the name
of the variable that is this struct, e.g. ''trainedModel''. \n \nThe
table, T, must contain the variables returned by: \n c.
RequiredVariables \nVariable formats (e.g. matrix/vector, datatype)
must match the original training data. \nAdditional variables are
ignored. \n \nFor more information, see <a href="matlab:helpview
(fullfile(docroot, ''stats'', ''stats.map''),
```

```
''appclassification_exportmodeltoworkspace'')">How to predict using
an exported model</a>.');

% Extract predictors and response
% This code processes the data into the right shape for training the
% model.
inputTable = trainingData;
predictorNames = {'tau1', 'tau2', 'tau3', 'tau4', 'p1', 'p2', 'p3',
'p4', 'g1', 'g2', 'g3', 'g4'};
predictors = inputTable(:, predictorNames);
response = inputTable.stabf;
isCategoricalPredictor = [false, false, false, false, false, false,
false, false, false, false, false, false];

% Perform cross-validation
partitionedModel = crossval(trainedClassifier.ClassificationSVM,
'KFold', 10);

% Compute validation predictions
[validationPredictions, validationScores] = kfoldPredict
(partitionedModel);

% Compute validation accuracy
validationAccuracy = 1 - kfoldLoss(partitionedModel, 'LossFun',
'ClassifError');
```

Cubic SVM

```
function [trainedClassifier, validationAccuracy] = trainClassifier
(trainingData)
inputTable = trainingData;
predictorNames = {'tau1', 'tau2', 'tau3', 'tau4', 'p1', 'p2', 'p3',
'p4', 'g1', 'g2', 'g3', 'g4'};
predictors = inputTable(:, predictorNames);
response = inputTable.stabf;
isCategoricalPredictor = [false, false, false, false, false, false,
false, false, false, false, false, false];

% Train a classifier
% This code specifies all the classifier options and trains the classifier.
classificationSVM = fitcsvm(...
  predictors, ...
  response, ...
  'KernelFunction', 'polynomial', ...
  'PolynomialOrder', 3, ...
  'KernelScale', 'auto', ...
  'BoxConstraint', 1, ...
```

```matlab
    'Standardize', true, ...
    'ClassNames', categorical({'stable'; 'unstable'})));

% Create the result struct with predict function
predictorExtractionFcn = @(t) t(:, predictorNames);
svmPredictFcn = @(x) predict(classificationSVM, x);
trainedClassifier.predictFcn = @(x) svmPredictFcn
(predictorExtractionFcn(x));

% Add additional fields to the result struct
trainedClassifier.RequiredVariables = {'tau1', 'tau2', 'tau3', 'tau4',
'p1', 'p2', 'p3', 'p4', 'g1', 'g2', 'g3', 'g4'};
trainedClassifier.ClassificationSVM = classificationSVM;
trainedClassifier.About = 'This struct is a trained model exported from
Classification Learner R2018a.';
trainedClassifier.HowToPredict = sprintf('To make predictions on a new
table, T, use: \n yfit = c.predictFcn(T) \nreplacing ''c'' with the name
of the variable that is this struct, e.g. ''trainedModel''. \n \nThe
table, T, must contain the variables returned by: \n c.
RequiredVariables \nVariable formats (e.g. matrix/vector, datatype)
must match the original training data. \nAdditional variables are
ignored. \n \nFor more information, see <a href="matlab:helpview
(fullfile(docroot, ''stats'', ''stats.map''),
''appclassification_exportmodeltoworkspace'')">How to predict using
an exported model</a>.');

% Extract predictors and response
% This code processes the data into the right shape for training the
% model.
inputTable = trainingData;
predictorNames = {'tau1', 'tau2', 'tau3', 'tau4', 'p1', 'p2', 'p3',
'p4', 'g1', 'g2', 'g3', 'g4'};
predictors = inputTable(:, predictorNames);
response = inputTable.stabf;
isCategoricalPredictor = [false, false, false, false, false, false,
false, false, false, false, false, false];

% Perform cross-validation
partitionedModel = crossval(trainedClassifier.ClassificationSVM,
'KFold', 10);

% Compute validation predictions
[validationPredictions, validationScores] = kfoldPredict
(partitionedModel);

% Compute validation accuracy
validationAccuracy = 1 - kfoldLoss(partitionedModel, 'LossFun',
'ClassifError');
```

Index

A

Advanced Message Queuing Protocol
  (AMQP), 164, 165
AlphaZero, 188
App engine, 162
Area under the curve (AUC), 150, 151
Artificial intelligence (AI) principles, 2
Artificial neural network (ANN)
  biological neurons category, 138
  deep learning, 139
  edge, 136
  human/animal brain, 136
  neurons, 136
  perceptron, 137
Asset aging
  challenges, classical models, 45, 46
  degradation (*see* Degradation process)
  hard physical and soft failure, 44
  life-cycle characters, 43
  lifetime data, 43
  measurements, 44
  PHM (*see* Prognosis and health
    management (PHM))
  RUL (*see* Remaining useful life (RUL))
  scheduled maintenance, 44
  working conditions, 43
Asset management (AM)
  CBM, 24
  data-driven algorithms, 26, 27
  definition, 21
  framework, 24–26
  implementation time, 24
  IoT
    human intervention, 36
    maintenance operators, 36

monitor, 35
    smart decision-making, 36
    smart devices and AI, 37
    smart environment, 36
    visibility, 35, 36
  life-cycle and supporting activity, 23
  maintenance activity, 21, 23
  maintenance strategy
    actions, 27
    CBM, 30
    corrective, 29
    data-driven approaches, 29
    deterministic approaches, 28, 29
    hybrid approach, 29, 31
    optimal approach, 31
    predictive approaches, 29
    preventive, 30
    priority, 27
    problem, 31
    stochastic approaches, 29
  performance indicators
    availability, 38
    capability, 39
    effectiveness, 39
    efficiency, 39
    maintainability, 38
    reliability, 38
    utilization, 39
  RUL (*see* Remaining useful lifetime (RUL))
  service activity, 23
  success, 22
  working conditions, 24
Azure cloud platform
  cloud gateway, 165, 167
  data analysis

CPSIA information can be obtained
at www.ICGtesting.com
Printed in the USA
LVHW080959270120
644899LV00001B/47